Force Control Theory and Method of Human Load Carrying Exoskeleton Suit

Zhiyong Yang · Wenjin Gu
Jing Zhang · Lihua Gui

Force Control Theory and Method of Human Load Carrying Exoskeleton Suit

国防工業出版社
National Defense Industry Press

Springer

Zhiyong Yang
Naval Aeronautical Engineering Institute
Yantai
China

Jing Zhang
Naval Aeronautical Engineering Institute
Yantai
China

Wenjin Gu
Naval Aeronautical Engineering Institute
Yantai
China

Lihua Gui
Naval Aeronautical Engineering Institute
Yantai
China

ISBN 978-3-662-54142-5 ISBN 978-3-662-54144-9 (eBook)
DOI 10.1007/978-3-662-54144-9

Jointly published with Original Chinese edition published by National Defense Industry Press, China, 2013

The print edition is not for sale in China Mainland. Customers from China Mainland please order the print book from: National Defense Industry Press.

Library of Congress Control Number: 2017930652

Printed on acid-free paper

This Springer imprint is published by Springer Nature
The registered company is Springer-Verlag GmbH Germany
The registered company address is: Heidelberger Platz 3, 14197 Berlin, Germany

Preface

The load capacity of human itself is limited. However, on many occasions, the human body needs to bear large load, such as earthquake relief, mountain land exploration, and building in mountains. In this case, only relying on the ability of human itself, problems often can not be solved, which has restricted human activities. In order to solve this problem, in addition to minimizing the weight and the size of load, we will naturally consider using external auxiliary devices to help people with heavy load to reduce human's energy consumption. Now, human has developed a lot of auxiliary devices, generally referred to as "exoskeleton", "power-assisted robot", "mechanical leg", "auxiliary leg", etc. Because the exoskeleton is similar to the invertebrates' exoskeleton that can be worn on human body like clothes, so it can be called "exoskeleton suit" very vividly.

The most remarkable characteristics of exoskeleton suit are wearable, operability, and intellectualization. The basic idea of its design is to combine human's intelligence with the strength of machinery: human sends control instruction to exoskeleton suit, and the exoskeleton suit provides the energy for the movement. The man–machine interaction intelligent device is different from the traditional robot, whose control system is controlled by human's consciousness, and by the complementation of people's intelligence and machinery's strength, not only can avoid the mechanical structure's complexity produced when the traditional robot simulates human's intelligence, but also can greatly expand the human body's ability. The ultimate goal of designing exoskeleton suit is to increase the wearer's speed, strength and endurance, and the wearer bearing heavy load can easily complete such actions as walking, running, jumping, squatting, and turning off, also the heaviness felt by the wearer being reduced significantly, the load of dozens of kilograms being felt like a few kilograms. The perfect coupling between human and exoskeleton suit conforms to the human body's engineering standard, safe and reliable, easy to operate and achieved the harmony and complementation of human and the machine. The appearance of exoskeleton suit is the innovation of transportation modes, which breaks through the limitation of traditional vehicles affected by terrain conditions easily. In the regions not suitable for marching of motor vehicles, such as rugged mountains and dense jungle, the exoskeleton suit can help people carry heavy load. Because it can significantly reduce the wearer's heaviness and save the wearer's physical power, so it can effectively improve human's mobility and continuous movement capacity; therefore, it will have a broad application prospect.

Although it has been for more than 50 years since the concept of exoskeleton suit was put forward and the birth of the first set of exoskeleton system, it is in the late 1990s that the exoskeleton system research was in full swing. Internationally, the United States, Japan, France, South Korea, Israel, and etc. have developed prototypes for demonstration, and many other countries have researched in some key technology areas of exoskeleton suit. Among them, the United States, Japan, and South Korea are most outstanding, whose technologies are most advanced, closest to the practical. Domestic study in exoskeleton suit began in about 2004, and Zhejiang University, Harbin Industrial University, Naval Aeronautical Engineering Institute, Soldier Systems Center of PLA General Logistics Department, and East China University of Science and Technology are the institutes that have started the research earlier. Then, Southeast University, Beijing Institute of Technology, Nanjing University of Science and Technology, Ordnance Industry Group, Tengzhong Aviation Electronics co. Ltd., and some other scientific research and teaching institutions have entered this field. Because the system involves the mechanical design, drive mode, control method, intelligent sense, power conversion, signal analysis, and some other disciplines and fields, and especially, the system is a complex man–machine coupling system, so it has played an important role in the teaching and scientific research. But due to a late start, compared with the research level of developed countries, we have been left behind.

Naval Aeronautical Engineering Academy in which the book's authors work is one of the institutes that have started the research earliest. And the authors have been in the research group, having a comprehensive understanding in the research of exoskeleton suit, and having got some certain research results on the research of exoskeleton suit's control methods. The publication of the book is desired to promote the development of the exoskeleton suit system.

This book introduces one of the key technologies of exoskeleton suit, namely the control method, and focuses on the force control theory and method based on multidimensional force/torque sensor. This book consists of eight chapters. Chapter 1, the introduction, introduces the concept and background of exoskeleton suit and analyzes the research status of exoskeleton suit home and abroad, the development status of exoskeleton suit's control method. Chapter 2 is the modeling basis of this book, in which the exoskeleton suit's mathematical models of kinematics and dynamics have been established. Chapter 3 introduces and analyzes the sensitivity amplification control methods of exoskeleton suit. From Chaps. 4 to 8, the force control methods of exoskeleton suit based on multidimensional force/torque sensors have been introduced. Among them, Chap. 4 analyzes the direct force control method of exoskeleton suit; Chap. 5 analyzes the force control method of exoskeleton suit based on the inner position loop; Chap. 6 analyzes the impedance control method of exoskeleton suit; Chap. 7 analyzes the impedance control method of exoskeleton suit with uncertainty; and Chap. 8 introduces the impedance control method of using neural network to estimate the exoskeleton suit's reference trajectory.

We would like to extend our sincere gratitude to Professor Lu Wen and Ma Hongxu of National University of Defense Technology, Professor Sun Fuchun of Tsinghua University, Professor Cao Heng of East China University of Science and Technology, and Professor Zhou Hong of Quartermaster Equipment Research Institute of the PLA General Logistics Department for their valuable advice and suggestions; we would like to extend our sincere gratitude to Professor Zhao Guorong, Huang Zhiyong, and Yang Xiuxia of Naval Aeronautical Engineering Institute for their guidance and help, and also we would like to extend our sincere gratitude to Zhu Yuguang, Li Shuangming, and Wu Guosheng for their contribution to the book. Sincerely thank all the people who have contributed to the publication of this book.

The research on exoskeleton suit system is an important and continuously developing research direction. The research of this book on exoskeleton suit concentrates on its force control technology, and for the comprehensive and practical research on the whole system, we still have a lot of work to do. We hope there are more scholars paying attention to the challenging research, having the related issues in this field studied further. Being constrained by authors' knowledge level, there must be many defects and mistakes in this book, and please do not spare your comments.

Yantai, China

Zhiyong Yang
Wenjin Gu
Jing Zhang
Lihua Gui

Contents

Symbols

DARPA	Defense Advanced Research Projects Agency
MIT	Massachusetts Institute of Technology
ZMP	Zero Moment Point
EMG	Electromyogram
GRF	Generalized Ground Reaction Force
GCP	Ground Contact Point
SAC	Sensitivity Amplification Control
RBF	Radial Basis Function Neural Network
m	Mass
I	Moment of Inertia
e_{ij}	The Unit Vector ^{i} Expressed in Coordinate System
q_i	Joint Angle
$\mathbf{p}$	Positon Vector of Operation Space
p_{x}	X Direction Poisition
p_{y}	Y Direction Poisition
θ	The Rotation Angle Around the Z Axis
$\mathbf{R}$	Rotation Matrix
$\mathbf{q}$	Joint Angle Vector
$\dot{\mathbf{q}}$	Joint Angular Velocity Vector
$\ddot{\mathbf{q}}$	Joint Angular Acceleration Vector
$\dot{\mathbf{p}}$	The Translational Velocity Vector of Operation Space
ω	The Rotation Angular Velocity Vector of Operation Space
$\mathbf{v}$	The Velocity Vector of Operation Space
$\mathbf{J}$	Jacobian Matrix
$\mathbf{x}$	The Position Vector of Generalized Coordinates
$\dot{\mathbf{x}}$	Generalized Velocity Vector
$\ddot{\mathbf{x}}$	Generalized Acceleration Vector
$\mathbf{x}_{\mathrm{e}}$	The Position Vector of Environment
$\mathbf{x}_{\mathrm{c}}$	The Orientation Vector of Reference Trajectory
$\dot{\mathbf{x}}_{\mathrm{c}}$	The Velocity Vector of Reference Trajectory

$\ddot{\mathbf{x}}_c$	The Acceleration Vector of Reference Trajectory
$\mathbf{x}_h$	The Position Vector of Human
$\mathbf{x}_d$	Desired Position Vector
$\mathbf{H}(\mathbf{q})$	Inertia Matrix
$\mathbf{C}(\mathbf{q}, \dot{\mathbf{q}})$	Coriolis Item
$\mathbf{G}(\mathbf{q})$	Gravity Term Matrix
$\mathbf{T}$	External Torque Vector
$\mathbf{a}$	Acceleration Vector
$\mathbf{f}_e$	The Acting Force Vector of Robot Acting on the Environment
$\mathbf{f}$	Human–machine Interaction Force Vector
f_x	Human–machine Force in X Direction
f_y	Human–machine Force in Y Direction
T_z	Human–machine Moment Around the Z Axis
$\mathbf{f}_d$	Desired Environment Acting Force Vector
$\mathbf{K}_{Pf}$	Spring Coefficient Matrix (The Stiffness Matrix)
$\mathbf{K}_{Df}$	Damping Coefficient Matrix
$\mathbf{K}_{Mf}$	Inertia Coefficient Matrix
$\mathbf{K}_{Pp}$	Desired Stiffness Coefficient Matrix
$\mathbf{K}_{Dp}$	Desired Damping Coefficient Matrix
$\mathbf{K}_{Mp}$	Desired Inertia Coefficient Matrix
$\mathbf{K}_P$	Proportionality Coefficient Matrix
$\mathbf{K}_I$	Integral Coefficient Matrix
$\mathbf{K}_D$	Damping Coefficient Matrix
$\mathbf{F}$	Movement Friction Coefficient Matrix
$\mathbf{T}_a$	The Torque/Moment Vector Exerted by Actuator
$\mathbf{T}_{hm}$	Human–machine Torque Vector
HF	Human Factor Matrix
HFM	The Maximum Human Factor Matrix
V	Lyapunov Function
G_a	The Kinetic Model of Exoskeleton Suit
G_h	Human–machine Interaction Model

Introduction

Exoskeleton suit can help human carrying heavy load, enhance their walking ability and velocity, and alleviate their fatigue after long march with heavy load. It not only has important application value in the military area, but also can be widely used in civilian areas of rescue, disaster relief, climbing, construction, assistive, etc. This book introduces one of the key technologies of the exoskeleton suit—control method. In addition to introducing the sensitivity amplification control, the book focuses on introducing the force control theory and method based on multiaxis force/torque sensor, which include the following: direct force control method, force control method based on inner position loop, impedance control method, impedance control with uncertainties, and reference trajectory estimate method of exoskeleton suit.

This book is mainly faced to those scientific research persons that engaged in the machine intelligent control theory, exoskeleton suit control, and design technology and provide valuable reference materials on the system design, integration, and implementation of exoskeleton suit. At the same time, the book is faced to those graduate students of the mechanical engineering, robotic technology, intelligent control technology, and control engineering and also can be used as a reference book of human–machine systems and technology.

Chapter 1
Introduction

1.1 Background

Human load capacity can refer to the load carriage standard for individual soldier. The individual soldier load is divided into combat load and marching load, mainly including weapons load, life load and single-soldier quartermaster equipment load. American military standard 1981 specified that the individual combat load should not exceed 35% of the soldier's body weight, the marching load should not exceed 45% of the weight, and for more than 95% of the soldiers, the combat load and the marching load are not more than 16.3 and 24.5 kg. According to the rule of Soviet army, the individual soldier load does not exceed one-third of body weight; the Soviet riflemen temperate combat load is about 17 kg, and the load in extraordinary cold regions is about 24.7 kg; the temperate marching load is about 23 kg, and the load in extraordinary cold regions is about 27.7 kg. Japanese self-defense forces think that soldier's average load should be around 15 kg; the temperate combat load is about 20 kg, and the load in cold regions is about 22 kg. For the Chinese People's Liberation Army, the combat load is 16 kg; under the general condition, the marching load is 25 kg; under the high-temperature environment, the marching load should be about 15 kg; and in plateau area, the load should be under 20 kg. However, in the actual war, because the soldiers are often required to carry as much ammunition and other equipment as possible, the individual soldier's actual load in many nation's army is far beyond the standard in the specification.

Take the American riflemen load in the Iraq war in 1991 as the example, as shown in Fig. 1.1, the complete set of American individual soldier's equipments includes camouflage bulletproof helmets and windbreak blindfold, desert camouflage uniforms and bulletproof vests, M4 rifle equipped with M68 sights (2 bullet belts containing 30 bullets), 1 grenade belt carrying 6 grenades, a musette bag (containing a set of chemical protective clothing, supplies of food, water and replenish ammunition, etc.), a first aid kit and multi-purpose bayonet, a flashlight, desert combat boots. These equipments make marching load reach 41.1 kg.

Z. Yang et al., *Force Control Theory and Method of Human Load Carrying Exoskeleton Suit*, DOI 10.1007/978-3-662-54144-9_1

Fig. 1.1 The American soldier with heavy load

Another example is the American army's "ground warriors" system, whose load is about 32.7 kg, and its shock-type structure can reduce weight 6.8–9 kg. These equipments as well as their necessary normal working power supplies become a heavy burden in "future soldier's" marching and combat. After the long march, the load can cause the damage of soldiers' back, waist and foot, which will influence the battle effectiveness.

It can be seen that the most of the load of soldiers are beyond the specified requirements. In order to make soldiers maintain good physiological status under the condition of load, every country has made great effort to do the research in individual soldier load, to develop the load-bearing device better conforming to human engineering principle. So various carried equipments for individual solider have emerged. By determining the carrying body area and carrying method properly, these carried equipments for individual soldier can ensure that the entire load is distributed on the human body evenly, reduce restrictions on human respiration and pulmonary function, and facilitate ventilation, which will improve the soldier's activity ability under the standard load. However, the carried equipment for individual soldier can only make the physical use more reasonable; it cannot substitute for people to carry heavy things, so it cannot help to solve the problem basically.

To solve this problem, in addition to using the means such as improving equipment material and processing technology to minimize the weight and size of the equipment, people naturally think of the idea that using external auxiliary devices to help load individual soldier's heavy weapons and equipments to reduce the physical consumption and preserve the battle effectiveness. Now, people have developed a lot of auxiliary devices, generally referred to as "exoskeleton," "power-assisted robot," "auxiliary leg," "mechanical leg," etc. These devices have generally rigid mechanism and externally worn like clothes by the human body, having the same effect as human skeleton of supporting load; therefore, it can be described as "exoskeleton suit" [1–3]. Exoskeleton suit is a new kind of mechanical device with wearable performance. From the perspective of the theory of intelligent system of human-machine integration, it is a human-machine integration system in which the human is the master and the machine is the auxiliary [4].

The most prominent features of the exoskeleton suit are its wearability, operability, and intelligence. The basic idea of the design is the combination of people's intelligence and mechanical strength that human gives exoskeleton control instruction, and the exoskeleton provides the energy for the movement; thus the tasks that human being cannot accomplish alone can be completed. Being different from traditional robots, the human-machine interaction intelligent device integrates the human body with bipedal walking robot together, making use of the human's intelligence to control the robot's walking which will simplify the autonomous robot's most common gait planning and gait stability problem and, at the same time, providing power to assist human's walking which will strengthen the ability and speed of human's walking and ease the human's fatigue that appears easily under the condition of high load and walking for a long time. The method of human's intelligence and mechanical strength complementing with each other can not only avoid the mechanism's complexity caused by the traditional robot simulating human intelligence but also can greatly expand the human's ability [5].

The ultimate goal of designing the exoskeleton suit is to increase the wearer's speed, strength, and endurance. The wearer can easily finish such action as walking, running, jumping, squatting, and turning off while bearing heavy weight; the wearer's weight-bearing feeling can be eased significantly, and dozens of kilograms of load is felt like a few kilograms. The perfect coupling between human and exoskeleton is conformed to human engineering standard, safe and reliable, operable, achieved the harmony and complementation between the human and machine.

The emergence of exoskeleton suit is an innovation of the way of transportation. It can quickly deliver large quantities of supplies and equipments, which breaks through the restrictions that the carriers such as the traditional vehicles are easily affected by terrain conditions. In rugged mountains, dense forests, and other areas that are not suitable for motor vehicles, the exoskeleton suit can help soldiers bear heavy weaponry. Since it can significantly ease the wearer's weight-bearing feeling and save the wearer's physical power, it will effectively improve the soldier's mobility and continuous fighting capability.

Predictably, the lower extremity exoskeleton will also have a very wide range of applications in civilian aspects. Take the disaster relief as the example, relief workers can use it to transport the rescue materials in the area that the vehicle cannot reach, and the firemen can use it to carry firefighting equipments up and down the stairs, etc. If the exoskeleton suit had been used in the Wenchuan earthquake relief work, how many people's lives would have been saved!

In abroad, the research of the exoskeleton suit has been going on for more than 40 years, but in our country, there are only a few institutes doing such researches after the year 2004, so there is a big gap between our country and international research level. Therefore, carrying out the research of the exoskeleton suit as soon as possible and developing our exoskeleton suit products with independent intelligent property rights have been imminent, which will have great significance in shortening the gap between our country and international level, also will effectively improve our army's single-soldier combat effectiveness.

1.2 Development Status of Exoskeleton Suit

For a long time, people have intended to make use of the machinery to increase the power of the human body, so as to extend human capabilities on the implementation level. Since the 1960s, many research institutions in the USA and the former Yugoslavia started the study of the power-assisted exoskeleton device almost at the same time. However, the focuses of the two countries' research are different. Scientists in the USA focused on developing a set of capacity enhancement device to improve the load capacity of normal people, mainly used in military field, while in the former Yugoslavia scientists focused on developing a power-assisted device to help the disabled [6, 7]. This book mainly discusses the previous exoskeleton suit, and the related content of the later one can be referred to the references [8–15].

1.2.1 Abroad Development of Exoskeleton Suit

1.2.1.1 The Exoskeleton Suit in the Early Stage

The research on the exoskeleton suit in early stage was mostly restricted to concepts, and the few research achievements with physical prototypes are not satisfactory. The first to design exoskeleton suit is Yagn from the USA, who has applied for several patents in 1890. Yagn's invention uses several longbow springs, which are parallel to human lower limbs to help the running and jumping of the human body. However, Yagn's design is purely mechanical, and there are not physical products and demonstration, as shown in Fig. 1.2 [8].

In 1963, Zaroodny, the scientist of the US Army Exterior Ballistics Laboratory, published a report, elaborating the program called "powered orthopedic supplement" he has worked on since 1951. His exoskeleton device was used to enhance the load capacity of healthy people (such as soldiers). This report is mainly about a concept study, emphasizing many fundamental difficulties, such as energy of portability, sensing and control, human-machine interface, and the exoskeleton's influence on human gait At the end of the report, it introduces a pneumatic prototype machine which might be the first power-assisted exoskeleton device driven by the external energy source. Although he did not receive further financial support for his design, his report has first expounded the engineering design of power-assisted exoskeleton device comprehensively [8].

From 1960 to 1971, the US GE Company was developing and testing a whole-body-shape exoskeleton device known as Hardiman; the purpose was to use it to augment the power of the human body itself, to help people finish the work of

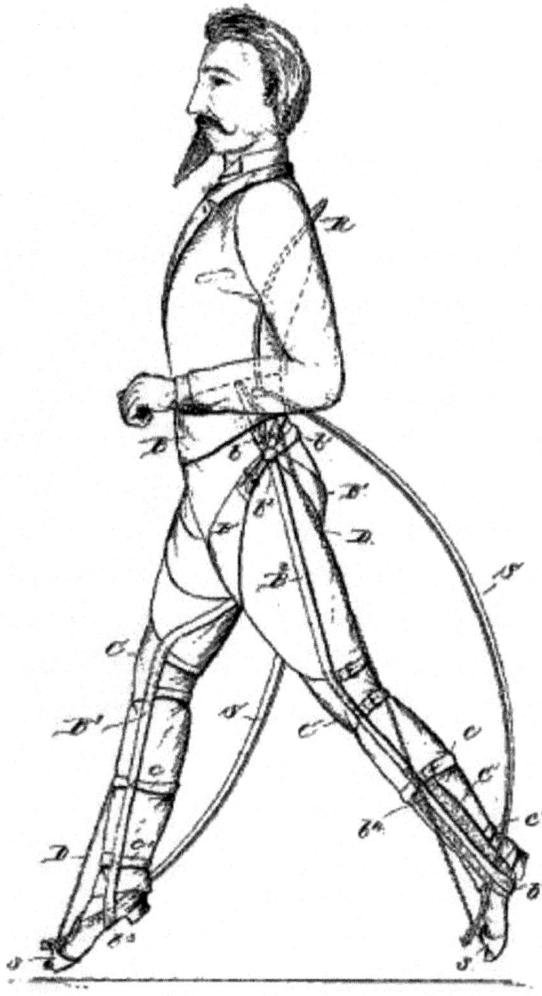

Fig. 1.2 Yagn's running auxiliary device

overload. Its large volume, weighing about 680 kg with 30 degree of freedom and adopting hydraulic actuator, can simultaneously enhance the operator's strength of upper and lower limbs. However, Hardiman only is able to augment an arm's power, and cannot enhance the capability of lower limbs and the whole body. Ultimately, due to security and many other factors, the Hardiman project was terminated. The most important contribution of the project is that many key issues in the design of the exoskeleton have been made clear, such as energy supply and human-machine interaction. At the same time, Hardiman project has delivered an important message that to develop a set of effective exoskeleton device is very difficult. Hardiman is shown in Fig. 1.3 [8, 16, 17].

Fig. 1.3 GE's Hardiman

1.2.1.2 The Exoskeleton Suit Project of DARPA

Since entering the twenty-first century, the main driving force of developing exoskeleton technology for energy enhancement is from Exoskeleton for Human Performance Augmentation (EHPA) project [18] belonged to the U.S. Defense Advance Research Projects Agency (DARPA).

The project was planned to develop a machine exoskeleton to improve human's military capabilities. The future soldier-wearing exoskeleton robot would become a super soldier with huge energy, who can carry more weaponry, and their fire power and protection level will be improved; at the same time, they can overcome obstacles, marching with high speed, so the fatigue caused by long march and heavy load can be reduced. The project was started in 2001, planned to begin the sampling test in 2005, and shifted to Program Executive Office Soldier (PEO Soldier) to carry out experiments in the army. The institutions or companies involved in the Projects include The Robot and Human Engineering Laboratory of University of California-Berkeley, Oak Ridge National Laboratory, Human Body's Function Institute in Salt Lake City, "Millennium Jet" Company, SARCOS Company, and Massachusetts Institute of Technology. Among them, the Berkeley

Branch Campus, SARCOS Company, and Massachusetts Institute of Technology have demonstrated the experimental prototype, while the other institutions have conducted the analysis and experiments [19, 20] in sensing, driving, and human-machine interface, biomechanics, human factor, testing, etc.

1. Berkeley Lower Extremity Exoskeleton (BLEEX) [5, 17, 21–35].

"Berkeley Lower Extremity Exoskeleton" (BLEEX) project, led by professor Kazerooni, made a successful presentation in 2004, as shown in Fig. 1.4. BLEEX's hip joints have three degrees of freedom (DOFs), its knee joints have one DOF, and ankle joints have three DOFs; there are four actuators: flexion and extension movement of hip joint, abduction and adduction of hip joint, flexion and extension movement of knee joint, and flexion and extension movement of ankle joint. Generally, there are energy storage springs [25] in the joints without actuators. The demand of BLEEX's kinematics and driving ability is based on the 75-kg men's walking medical gait analysis data [23, 25].

BLEEX is driven by bidirectional hydraulic cylinders which are installed on the rotary joints with triangular configuration way, making the actuating arm change with the joint angle. The BLEEX consumes 1143 W hydraulic energy and 200 W electric energy while walking on the flat ground. Correspondingly, a 75 kg of

Fig. 1.4 Berkeley's BLEXX

people walking on flat ground consumes 165 W metabolism energy; the reason why hydraulic drive was chosen in BLEEX is that its energy to weight ratio is the largest, so the volume is the smallest [25]. However, further studies have shown that compared with the hydraulic pressure drive, the motor drive can greatly reduce the energy consumption while walking on the flat ground, even though a joint motor drive is two times higher than the hydraulic joints in weight [29, 33].

BLEEX project adopts the sensitivity amplification control method [17, 21, 22, 24, 34] and develops a set of electrohydraulic portable energy supply device [30]. A 75-kg person who wears the BLEEX can walk easily at the speed of 1.3 m/s with a 34-kg backpack. BLEEX project addresses several key technologies of exoskeleton design, including mechanical structure, control strategy, and energy supply, and it can be said that BLEEX is the first real intelligent load-carrying exoskeleton that can be applied practically.

2. The SARCOS exoskeleton [8, 36].

SARCOS Company located in Salt Lake City has developed a whole-body-shape "Wearable Energetically Autonomous Robot (WEAR)" under the EHPA framework, as shown in Fig. 1.5. Just like its name, the WEAR is self-energizing, carrying its own power source. Similar to BLEEX, WEAR also adopts hydraulic drive, and the difference is that WEAR uses rotary hydraulic actuator directly to drive the joint movement. SARCOS is committed to developing a set of energy device to provide energy for WEAR's hydraulic servo valve.

SARCOS has demonstrated its exoskeleton prototype successfully and has achieved its many important functions: It can carry 84-kg load; the operator can carry one person on the back standing on one leg and can march at the speed of 1.6 m/s with 68-kg load on the back and 23-kg load in hands; and the prototype can get through 23-cm-deep swamp and can finish such actions as flexion over, squatting, and kneeling. After the EHPA project, SARCOS won the PEO Soldier's further substantial funding to develop their exoskeleton suit into personal combat vehicle (PCV).

3. MIT exoskeleton suit [8, 37–40].

The Media Laboratory's Biomechatronics Group of Massachusetts Institute of Technology (MIT) proposed a concept of quasi-passive exoskeleton control in the second phase of EHPA project funded by DARPA. As shown in Fig. 1.6, the goal of this concept was to use the human body's passive dynamic walking to develop a lighter, more efficient exoskeleton device. Passive dynamic walking refers to the robot walking completely dependent on gravity without external energy, first put forward by McGeer in 1989. The concept, passive dynamic walking, would bring a new solution to the robot's energy consumption problem [41–43]. However, the passive bipedal walking robot has no power, so it can only walk down a small slope, and the movement energy is actually transformed from the gravitational potential energy. In order to make the passive dynamic bipedal robot be able to walk on the flat ground, a small amount of driving force must be provided to its

Fig. 1.5 Exoskeleton suit by SARCOS

corresponding joints (hip or ankle) to start walking and compensate for the loss of energy during walking, which is called a quasi-passive dynamic.

MIT adopts the concept of quasi-passivity to design the exoskeleton suit, and its joints do not need any driving device instead of using energy stored in the spring in a negative energy phase of the walking gait to provide auxiliary power. By the dynamics and kinematics analysis of human body's walking movement, select the quasi-passive components (such as spring and variable dampers). The MIT's exoskeleton suit is coupled with the body with the shoulder strap, the waistband, the thigh brace, and special shoes. When there is no load, the exoskeleton suit weighs 11.7 kg, and only 2 W electric energy is needed to control the magnetorheological damper installed on knee joints. Experimental results show that the exoskeleton suit based on quasi-passive dynamics can successfully bear 36-kg load and walk at the speed of 1 m/s. In the single support phase, the exoskeleton suit transfers 80% of its

Fig. 1.6 Exoskeleton suit by MIT

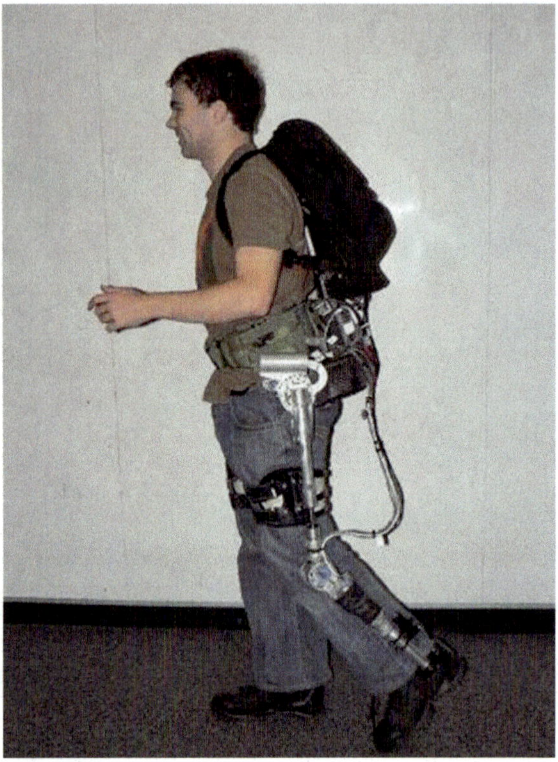

load to the ground. However, experimental results of MIT exoskeleton suit's metabolism are disappointing. For the same 36-kg load, the metabolism when using the exoskeleton suit is 10% more than without using exoskeleton suit.

4. Other key technologies

Under the support of the DARPA's exoskeleton suit project, many other key technologies also get a breakthrough, which can be used as the basis for further research, mainly including the following: the energy device based on monopropellant developed by the Berkeley [30, 44, 45], the way of exoskeleton motor drive [29], the compound control method of sensitivity amplification control and master-slave mode [24], high-speed communications network [34, 46, 47], etc.; the sole force/torque sensor, control strategy, energy technology, etc. developed by Oak Ridge National Laboratory [19, 20]; the pressure-sensing insoles used for the exoskeleton suit's gait recognition developed by Sandia National Laboratory, adopting micro-mechanical technology [48]; pneumatic lower extremity muscle intensifier put forward by the American Rutgers University [49]; and the analysis and research in different aspects of biomechanics by several institutes [23, 25].

1.2.1.3 Other American Exoskeleton Suit Projects

1. Lockheed Martin's Lower Extremity Exoskeleton (HULC)

Although BLEEX did not get further funding from DARPA, professor Kazerooni and his students founded Berkeley Bionics Company, striving for attracting venture capital and making marketization operation for the technology. The company developed a lighter, concise, and practical HULC (Human Universal Load Carrier) [18, 50], as shown in Fig. 1.7. HULC was purchased by famous weapon contractor, Lockheed Martin. The HULC weighs 24 kg (excluding batteries), and the two batteries weigh 3.6 kg. If soldiers wear HULC, they are able to bear 91-kg extra load, which is three times of the load capacity of BLEEX system. The battery can power for walking for 3 h continuously at the speed of 5 km/h, and the peak speed can reach 16 km/h. It can be said that the HULC is the closest to the

Fig. 1.7 HULC by Lockheed Martin

actual combat application of the exoskeleton suit and is currently being under further integrated development and the demonstration test in the troops.

2. Berkeley Bionics's Lower Extremity Exoskeleton (eLEGS)

Professor Kazerooni also developed the civilian exoskeleton suit. They developed the eLEGS (Exoskeleton Lower Extremity Gait System) mainly used for people's rehabilitation and assistance who suffered from lower limb paralysis, as shown in Fig. 1.8. Based on natural gait data, the researchers collected data on walking people's joint angle, knee flexion degree, toe flexion degree, and so on and then put them back on the eLEGS. In this way, the eLEGS will be conformed to people's habits better, and the body posture monitoring system can let it judge

Fig. 1.8 The eLEGS of Berkeley Bionics

people's intentions also help people to take the next step. Meanwhile, the sensor on the heel and sole will ensure that the feet will not leave the ground at the same time to avoid losing balance. Now, eLEGS is still in the experimental stage, but its performance has been very excellent. It is only 45 lb powered by the battery which can last 6 h [51].

3. Raytheon's exoskeleton [8, 36].

After Lockheed Martin had purchased the HULC, people more firmly believed that the exoskeleton suit would have a bright future. Immediately, weaponry giant Raytheon acquired SARCOS Company which had participated in the EHPA project and renamed SARCOS's first-generation exoskeleton suit to XOS. In 2010, XOS-2 came out, and its energy consumption has been reduced by half, much faster and stronger than the first generation, as shown in Fig. 1.9. Although XOS-2 still cannot leave the ground to get power supply, but its energy consumption has already been half of XOS-1. We believe that with the development of technology, the technical

Fig. 1.9 Raytheon's exoskeleton suit

problem of small energy module should be able to be resolved, when the exoskeleton suit will enter the rapid development period. It is a pity that the various technical information about the XOS has not been publicized until now [52].

1.2.1.4 The Exoskeleton Suit Projects in Japan

1. Hybrid Assistive Leg (HAL) of Tsukuba University [53–61].

At present, the high-performance exoskeleton that can be used for both energy aid and assisting the disabled person's walking at the same time is the world's first commercial exoskeleton robot "Hybrid Assistive Leg" (HAL) developed by Yoshiyuki Sankai and his team of Tsukuba University, and the model HAL-3 and model HAL-5 of this robot are more famous, as shown in Figs. 1.10 and 1.11.

Compared with HAL-3, the HAL-5 is more beautiful and portable: the joint motor is shorter and smaller, the energy supply device is smaller and its service time is longer (about 160 min), and more important is that the upper limb power-assisted function has been added. HAL's movement is completely controlled by the user with the automatic controller, which does not require any console or external

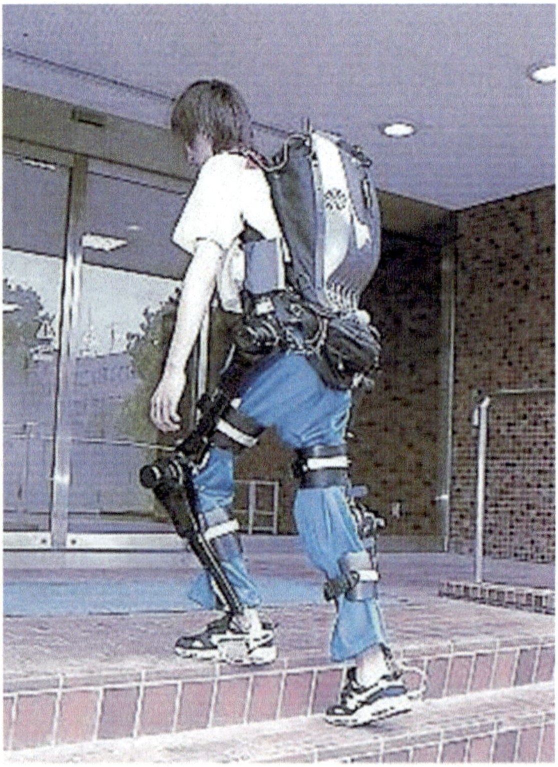

Fig. 1.10 The HAL-3 of Tsukuba University (the year 2004)

Fig. 1.11 The HAL-5 of Tsukuba University (the year 2006)

control. HAL can help the disabled walk at the speed of 4 km per hour, and easily climb the stairs. HAL composes of knapsack, built-in computer, batteries, inductive control equipment, and motor drive devices. The DC motor and harmonic gear reducer are installed on HAL's joints directly to drive each of its joints.

HAL is coupled with the human body through special shoes (mounted with a ground reaction force sensor), the brace belt of thigh and lower legs, and the large waist belt. It adopts surface electromyography sensor (EMG); the ground reaction force sensor; the potentiometer used for measuring the joint angle, gyroscope; and accelerometer installed in the back rack used to estimate the trunk posture, etc., and it is also equipped with portable energy systems. HAL is worn by healthy people for being displayed, and the further research on the application to the disabled has been carried on. But the video that can be found just shows HAL's upper limb auxiliary functions, as shown in Fig. 1.11: HAL's upper limbs can help the operator hold a weight of 40 kg. HAL's auxiliary functions for lower limbs have not been shown, although the inventor declared that HAL could help the bearing capacity of the users' lower limbs increasing from 100 to 180 kg. Although HAL can be used for energy auxiliary, because the exoskeleton suit's market prospect in the field of energy auxiliary is unclear; at present, the main research direction of professor Sankai is HAL's upper limb assistance functions and its applications in helping the disabled [62]. The professor has set up the Cyberdyne Company for HAL's marketing promotion, and a branch has been founded in Norway.

2. The power-assisted exoskeleton suit of Kanagawa Institute of Science and Technology [63, 64].

In Japan's Kanagawa Institute of Technology, scientists have developed a "Wearable Power Suit" (WPS), as shown in Fig. 1.12. Its lower limbs adopt a unique pneumatic drive to control the flexion and extension of the hip and knee joints, and the air pressure is supplied by a small pump installed on each actuator. With flexible belts, WPS installs force sensing resistors (FSRs) on the muscle

Fig. 1.12 The power-assisted exoskeleton suit of Kanagawa Institute of Science and Technology

surface (thigh rectus controlling the movement of knee joints) of operators, forming the muscle hardness sensors. When knee joints are bent and muscle is contracted, the stress on the FSRs increases, which is used to determine the operator's movement intensions together with joint angles obtained by the potentiometer measurement, so as to determine the needed joint torque. The WPS is specially developed for nurses, to help them take care of the patients whose weight is large or who are unable to walk, and experiments show that it can have people's strength increased by half to one times. WPS's mechanical design is very distinctive: there is no mechanical device in front of the operator, which makes caregivers contact with their nursing objects directly, ensuring the patients' comfort and safety.

3. Honda Corporation's power-assisted mechanical leg

Honda Research Institute has developed a walking booster based on HAL-3 which was developed by Honda Research Institute and Tsukuba University together, as shown in Fig. 1.13. Structurally, this newly developed wearable walking

Fig. 1.13 Honda corporation's exoskeleton suit

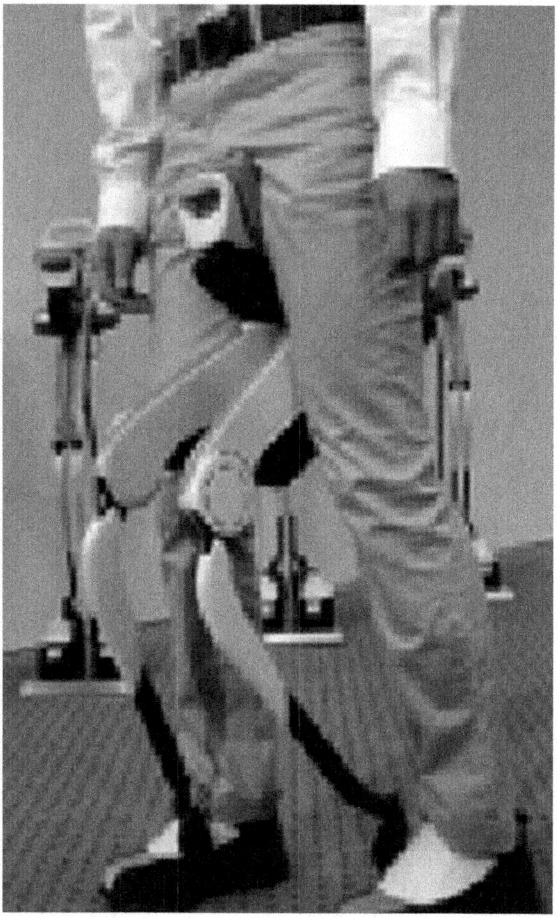

aid tool displayed by Japan's Honda, whose mechanical frame is connected with shoes, is installed a saddle seat in the upper part, and its double lower limbs have six DOF (degree of freedom), namely each leg's hip joint, knee joint, and ankle joint have one DOF. The structure adopts non-humanoid design, no connection being bound to the human body, the supporting seat bearing human body's weight, knee joint, and hip joint being used as the driving joint. Currently, the booster has been sold as the aid-tool product and has been applied practically. It not only can be used to help the elderly to walk and also can be used to help labors to improve work efficiency in certain kinds of work.

4. The exoskeletal robot of Tokyo University

Researchers, from Agriculture and Technology School of Tokyo University, have developed an exoskeleton robot system, as shown in Fig. 1.14. This system can provide powerful external forces for human users, which can be used to improve farmers and gardeners' work; that is to say, this product is mainly developed for agriculture area. Its total weight is about 20 kg and also can support its own weight. The product contains eight motors and 16 sensors totally, and these important components can help the wearer's strength increase instantaneously. Currently, the system is still in the stage of research.

Fig. 1.14 The power-assisted suit of Agriculture and Technology School of Tokyo University

1.2.1.5 The Exoskeleton Suit Projects in South Korea

1. The exoskeleton suit of Hanyang University

Hanyang University developed a multisensor lower limb movement information acquisition and gait analysis system for acquiring human lower limb movement information and analyzed the gait, as shown in Fig. 1.15. The system includes three parts: servo leg system includes three rotational joints (exoskeleton joint with one DOF, knee joint with one DOF, and ankle joint with one DOF) on each leg, and corresponding to the degree of freedom of leg rotation, six photoelectric encoders, 1 pair of FSR (foot sensor resistors) dynamometer shoes (installed with four pressure sensors on the sole of each shoe, with one sensor on the heel, on the front hallux

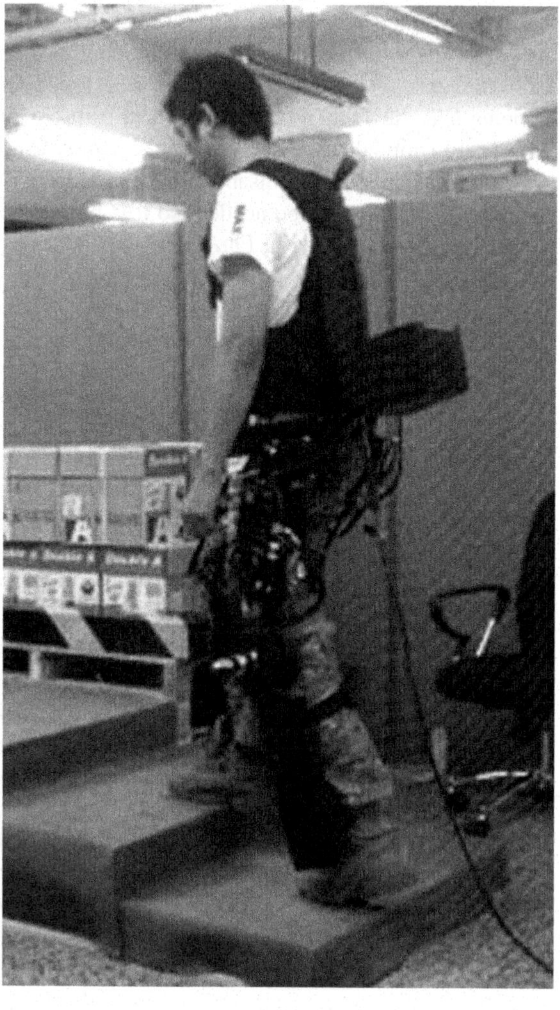

Fig. 1.15 The exoskeleton suit of Hanyang University

toe, and on the left and right side of the sole, respectively) are mounted with; data acquisition and processing system; and the gait analysis software. The system can obtain the joint movement information of the servo leg in real time, and it also can analyze the gait characteristics of the wearer in walking, single leg swinging, and sitting and standing movement. In addition, in terms of human lower limb motion analysis, this system proposes a gait analysis method of comprehensive servo leg system joint movement information and dynamometer shoes pressure information [65, 66].

2. The exoskeleton suit of Sogang University

After years of research in exoskeleton suit, Sogang University has achieved great achievements, publishing pieces of high level papers in the journal of IEEE's Electromechanical Integration, stating their research results in exoskeleton suit, mainly including the rotating elastic series actuator's application in the human-machine interaction robot [67], the gait monitoring system based on the air pressure sensor [68], the exoskeleton suit control method based on virtual gain [69], etc. The exoskeleton suit they have developed includes EXPOS (Exoskeleton Patients and the Old by Sogang) and its subsequent prototype SUBAR (Sogang University Biomedical Assistive Robot), mainly used for rehabilitation therapy and walking assistance of the disabled and the elderly. Its disadvantage is that a large auxiliary station should be needed, so it is not suitable for such occasions as labor assistance, etc. as shown in Fig. 1.16.

1.2.1.6 Other Exoskeleton Suit Projects

1. The exoskeleton suit of Nanyang Technology University [70–73].

Nanyang Technology University in Singapore is also doing the research and development in lower extremity exoskeleton, as shown in Fig. 1.17, called NTULEE (Nanyang Technology University Lower Extremity Exoskeleton). NTULEE is composed of internal and external exoskeletons: internal exoskeleton is simple and small, installed directly on the operator, and encoders on the joints are used to measure angle information of the human body's joints; the external exoskeleton mainly bears load on the back. After the conversion of algorithm, the internal exoskeleton measurement information forms control signal, and then the control signal is transformed to control the external exoskeleton movement. The existence of both internal and external exoskeleton ensures the exoskeleton suit tracks human body movement, and at the same time, they have designed shoes mounted with two-layer force sensors to measure the zero-moment point (ZMP) of human body and exoskeleton suit, respectively, making use of the control signal generated from the deviation between the human body and skeleton suit's ZMP to control the exoskeleton suit's ZMP tracking human body's ZMP, so as to guarantee the exoskeleton suit's stability in marching. They use the xPC MATLAB target

Fig. 1.16 The exoskeleton suit of Sogang University

toolbox as a real-time operating system and plan the control behavior of the whole system according to the commonly used four layers of biped walking robots.

2. RoboKnee [74].

In 2000, four students graduated from MIT leg laboratory in Cincinnati set up a company, specializing in robot research. They developed a simple exoskeleton (RoboKnee) used to enhance the strength of the knee joint, so as to make the wearer

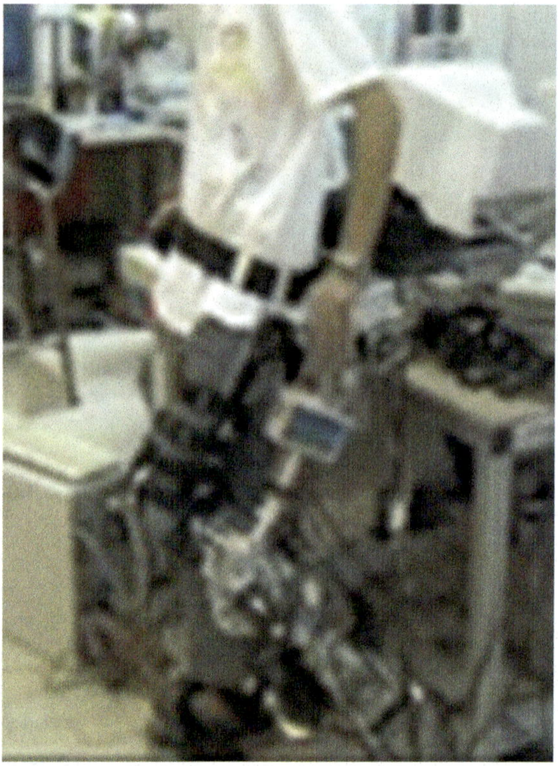

Fig. 1.17 The exoskeleton suit of Nanyang Technology University

can complete the actions of going up and down the stairs and squatting easily while bearing load, as shown in Fig. 1.18. RoboKnee uses series elastic actuator (SEA) developed by themselves together with thigh and shank's brace belts to link the upside and downside parts of the knee joint. Within the normal leg speed, RoboKnee's impedance is very low. There is no control panel, switch, rocker, display, or other user interface in the RoboKnee. The operator can complete the actions of going up and down the stairs and squatting deeply, etc. without needing any training. Using 4-kg NI-MH battery, RoboKnee can work for 30–60 min, and if using silver-zinc battery, the working time will double, but it still cannot meet daily work requirement. RoboKnee does not pass the load directly to the ground by itself, but using lower limbs of the human body to complete this task; therefore, RoboKnee cannot enhance human body's load capacity (still limited to the bearing capacity of human body's skeleton). In other words, RoboKnee can only enhance the muscle's strength, but cannot enhance the skeleton's strength. Because the actuator is installed in the back of the thigh, so the wearer cannot sit down. RoboKnee is just a simple verification device, without hip and knee joints' power assistance function, so still there are many functions needed to be improved.

Fig. 1.18 Yobotics's RoboKnee

3. Israel's "ReWalk".

"ReWalk" is a set of lower limb power-assisted exoskeleton used for medical rehabilitation, whose research is led by Amit Goffer of Israel's Argo Medical Technologies Company. It can help paraplegia patients (people who are paralyzed blow the waist) stand, walk, and climb stairs. As shown in Fig. 1.19, "ReWalk" consists of electric leg support, body sensors and a backpack, and a pair of crutches to help maintain the body balance. There is a computer control system and an accumulator built in the backpack. The user can first use remote control belt to choose a certain setting, such as standing, sitting, walking, climbing and so on, then leaning the body forward, so the body sensors can be activated to start the device.

4. Russian exoskeleton suit development plan

On October 27, 2009, the Third Central Research Institute of Russian Defense Ministry announced that they would develop the "Boyets 21" combat exoskeleton suit successfully before 2015. The suit allows infantry to run rapidly with heavy load and can be took off quickly when the power is exhausted. The suit is 14 kg lighter than that of the last generation. Russia did not release the suit's load capacity. In addition, there is news showing that Moscow Engineering Physics

Fig. 1.19 The ReWalk of Argo Medical Technologies Company

Institute is studying using artificial muscles (nanomaterials whose length changes with the electromagnetic conditions) instead of the hydraulic system to drive the exoskeleton robot. The development target of the artificial muscle fiber is that making use of voltage change causes the fiber's expanding or contracting which can produce power, so the hydraulic components can be replaced. This can further reduce the energy consumption of the exoskeleton robot, and the system with fewer moving parts is more reliable. Moreover, processing wire flexion problem is easier than processing hydraulic pipeline flexion, and there is no liquid leakage problem. So, in the future, the mainstream exoskeleton robots may not continue to use the hydraulic system.

5. The other exoskeleton suits that have not been demonstrated

There are many other exoskeleton suit research projects demonstrated publicly, and there is little relevant information, but the limited information also has certain reference value, such as the researchers of Italian PERCRO Laboratory adopt new compact portable actuator to design exoskeleton suit [75]; Japanese Hokkaido University designs a exoskeleton suit mainly used for moving heavy weight, adopting the EMG sensor similar to HAL [76, 77]; the exoskeleton suit's research of Japanese Mie University specializes in the power assistance mode of lifting weight [78]; Italian Biological Robot Institute designs a pressure-sensing insole to perceive foot insole information of human walking, which can be used to judge the human body's walking model in the exoskeleton suit [79]; American Northwestern University has developed a negative damping control (the special case of active impedance control) method of single degree of freedom exoskeleton suit [80]; Italian Institute of Technology has developed a new type of variable stiffness actuator and has applied it to human-machine interaction robot that is similar to the exoskeleton suit [81]; Sydney University of Science and Technology has done the preliminary research of lower limb exoskeleton suit power system of single degree of freedom [82]; California Ivy League College has designed two kinds of exoskeleton suits with the pneumatic artificial muscle, one for assisting knee movement and the other for assisting finger movement [83].

1.2.2 Domestic Development Status of Exoskeleton Suit

Domestic research in exoskeleton suit is relatively lagging behind, and there has been only a small amount of literatures involving the study of exoskeleton suit after 2004, mostly in theory research.

The following institutes have demonstrated prototypes:

1. Zhejiang University

The Machinery Electronic Control Engineering Research Institute of Zhejiang University is doing the research. They have designed a set of exoskeleton prototype test system using pneumatic system, based on hierarchical control framework of human-machine coupling, as shown in Fig. 1.20 [84], and have designed exoskeleton suit's control system in detail [85].

2. University of Science and Technology of China

University of Science and Technology of China has done the analysis and research in the configuration, perception and control methods of exoskeleton suit, and the prototype is shown in Fig. 1.21 [86]. In recent years, they have studied the exoskeleton suit's perceptive system deeply [87].

3. East China University of Science and Technology

East China University of Science and Technology has studied in the exoskeleton suit's virtual prototype and has developed the physical prototype. Its principle is similar to the HULC, adopting the hydraulic drive method, and a variety of functions of the plantar pressure-sensing and joint power assistance have been realized, as shown in Fig. 1.22 [88–91].

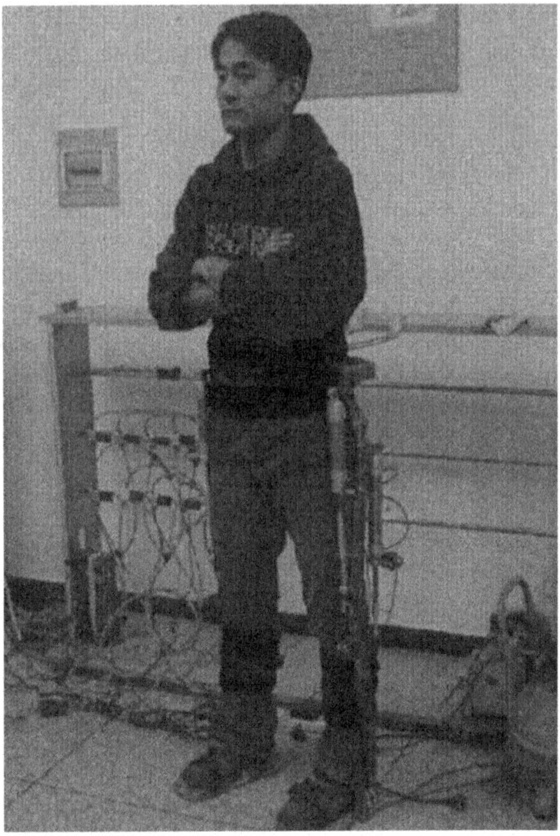

Fig. 1.20 The exoskeleton suit prototype of Zhejiang University

4. Naval Aeronautical Engineering Institute

Naval Aeronautical Engineering Institute is one of the institutes that has studied the exoskeleton suit earliest in China. They completed the first exoskeleton suit in 2006, which was named the NES-1 (The First-Generation Naval Aeronautical Engineering Institute Exoskeleton Suit). The prototype adopts the motor drive mode, crafty using gas spring to realize joint power assistance, showing walking and other functions. However, the gait pattern is limited, as shown in Fig. 1.23.

The second generation of the exoskeleton suit prototype NES-2 developed by Naval Aeronautical Engineering Institute came out in 2008, as shown in Fig. 1.24. This prototype adopts new design idea: the drive motor is not installed directly in the knee joint, but with the flexible inhaul cable device, isolating the electric motor and joints in space; thus, the weight on the leg of the exoskeleton suit is greatly reduced, and the motor and its actuator is installed as a part of the load on the back rack. This structure ensures that the swinging legs need not be controlled, which simplifies the control method.

In 2009, the institute designed the third-generation exoskeleton suit prototype NES-3, whose waist ring, sensing boots, joint design, structure design, etc. have

Fig. 1.21 The power-assisted robot prototype of University of Science and Technology of China

been further improved, as shown in Fig. 1.25. The followings are the NE-3 prototype performance index:

(1) Prototype weight: 18.5 kg;
(2) the maximum load bearing: more than 40 kg;
(3) the normal working load: 30 kg;
(4) the normal working pace: about 1 m/s;
(5) the continuous working time: about 2 h; and
(6) the gait tracking: walking, swiveling, going up and down the stairs, squatting, kneeling down, sitting, side kicking, crawling, going up and down the slope, and other actions. As shown in Fig. 1.26.

There are few reports on the exoskeleton suit in China, and until after 2004, only a small amount of literatures involving exoskeleton suit have appeared. Zhejiang University has first done some research in the suit's remote teleportation system, but its emphasis lies in teleportation, so its essence is a master-slave control mode, which is far from the idea of exoskeleton suit [92]. Then, Zhejiang University has done some study in the exoskeleton's movement mechanism based on the mechanical design and control method and has developed a prototype, but there has

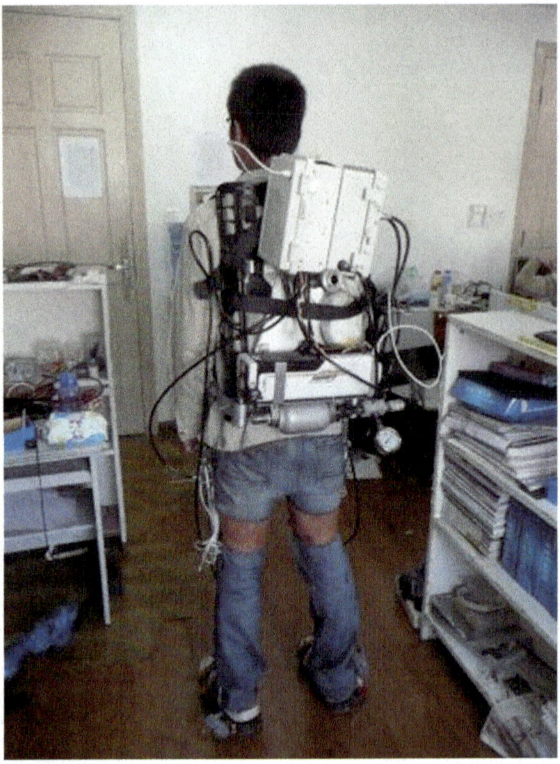

Fig. 1.22 The exoskeleton suit prototype of East China University of Science and Technology

been no public data showing it has fully realized walking power assistance, and the prototype is shown in Fig. 1.20 [84]. Harbin Institute of Technology has studied the power-assisted mechanical leg used for helping lower limb disabled patients [93] and then get the human motion data using image method, while conducting the exoskeleton suit's kinematics simulation using MATLAB [94]. University of Science and Technology of China has analyzed and researched the configuration, perception and control methods of exoskeleton suit, and the prototype is shown in Fig. 1.21 [86, 95, 96]. East China University of Science and Technology has studied the exoskeleton's virtual prototype and has developed physical prototypes, as shown in Fig. 1.22 [88, 89]. Nanjing University of Science and Technology has done some researches in the working mechanism of the exoskeleton suit using the virtual prototype technology [97]. Southeast University has analyzed the exoskeleton's kinematics and has designed virtual prototypes, but just in their infancy [98]. University of Hong Kong has researched the applying method of magnetorheological body drive in exoskeleton suit [99]. Overall, the domestic research in exoskeleton suit started much late, and until now, the fully functionalized prototype has not come out. Therefore, speeding up the research of all kinds of technologies of the exoskeleton suit is of great significance to improve the exoskeleton's research level in China and fill the domestic technology blank.

Fig. 1.23 The prototype of NES-1 of Naval Aeronautical Engineering Institute

1.3 The State of the Art of the Research in Exoskeleton Suit Control Methods

The biggest difference between the exoskeleton suit and other robots is that its operator is not a machine but a person, who is in the loop, namely "Man in Loop." There is a real physical contact between the operator and exoskeleton suit, forming an integration system of human-machine coupling. The control purpose of human-machine coupling system is to make human and machines work coordinating to complete tasks. People in the loop complete the environmental awareness and behavior decision tasks including deciding movement patterns, movement direction, speed, and the exoskeleton suit needs to judge the human's intentions through the sensor so as to follow the movements of the human body. The study of exoskeleton suit system involves the ergonomics, biology, machinery, electronics, automatic control, computing, and many other subjects. This book mainly studies the control method of the exoskeleton suit, and the following paragraphs will analyze the control methods that have been used in the exoskeleton suit and that will be used possibly, so as to provide guidance for the design of the exoskeleton suit's controller.

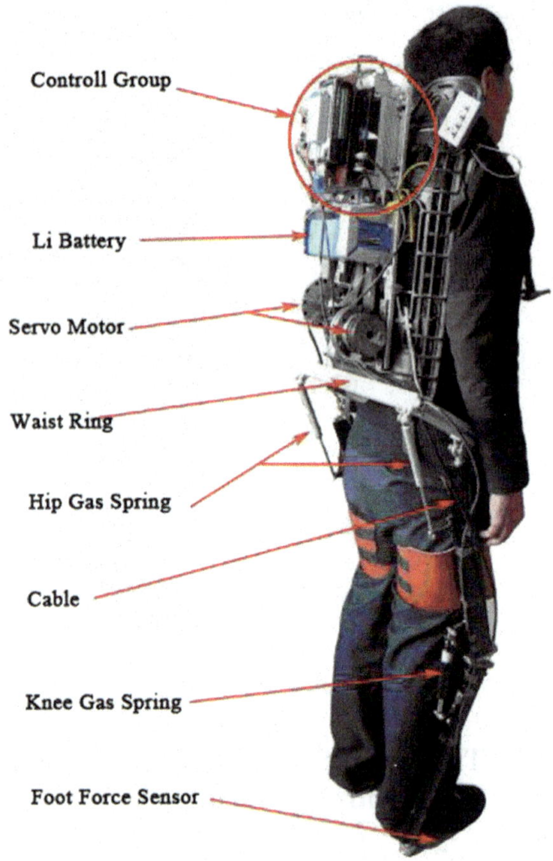

Fig. 1.24 The prototype of NES-2 of Naval Aeronautical Engineering Institute

1.3.1 Operator Control

Relying on the operator's judgment on the environment, themselves, and the exoskeleton suit motion state, controlling the exoskeleton suit movement with appropriate switch, button, and other devices is a very visualized idea, and even the operator of the exoskeleton suit need not be the wearer himself. A lot of gait rectification devices with external energy actuators for the purpose of rehabilitation in the exoskeleton suit are based on this method [21]. These rectification devices are mainly lower limb exoskeleton used to support the weight and do the lower limb rehabilitation training for operators. These devices' command signals generally come from healthy bodies. For example, the exoskeleton suit designed by Yano has a switch and ground reaction force sensor, driving hip joints' movement by the operator's control to change the distance of the feet soles leaving the ground [100]. Johnson has designed a kind of movement auxiliary device for the paralysis,

Fig. 1.25 The prototype of NES-3 of Naval Aeronautical Engineering Institute

amputation, and lower limb paralysis patients, and it controls the movement of each leg by finger joint sensors. Finger movement is monitored by 6–8 goniometers, finding a pair of one-to-one mapping between the finger joint movement and the corresponding leg joints, and getting other controlling characteristics by a set of switches installed in the hand. The first generation of prototype of Naval Aeronautical Engineering Institute uses the motion of the forearms to control the motion of legs and also has realized the function of bearing load [101]. The disadvantage of this approach is that no matter which part of the body used by the operator to release the order, once the part is determined, it will not be able to perform other activities, and moreover, the operator must command continuously, not only wasting the energy but also affecting the normal movement of the operators.

Fig. 1.26 The movement modes accomplished by NES-3

1.3.2 Preprogrammed Control

Some exoskeleton devices run by preprogramming, and the operator can only do limited intervention, such as "stop" or "start" [21]. These devices are generally rehabilitation devices designed for lower limb disability patients, whose trajectory has been programmed in advance; the design is based on the movement of normal gait and some changes have been made to adapt to the orthotic device. When running, the exoskeleton suit drives patients' movement to help patients do the recovery training of walking ability. The followings are some examples of orthotic devices: Ruthenberg's active gait orthotic device [102]; Vukobratovic's rehabilitation device [6, 7]; Colombo's gait correction device, etc. [9]. All these rehabilitation correction devices based on program control require patients to use walking stick or additional auxiliary framework to maintain the walking stability of operators, and the movement forms are also very limited.

1.3.3 Myoelectricity Control

In 1851, French scientist Dubois-Reymond first put forward the muscle sensing. Nowadays, electromyograms (EMGs) model has been developed from linear model to nonlinear model, widely applied to the design and usage of myoelectricity sensors. Human muscle's activity can be measured through surface electrode or implanted electrode to get the electromyograms and through which the muscle's excitement degree can be judged to estimate human's movement intentions. There are many kinds of rehabilitation mechanism controlled by electromyograms: some control arms [103], some control fingers [104, 105], some control lower extremities [106, 107], and some control other parts. And one of the most successful exoskeleton projects that have applied EMGs to strengthen the lower limbs' ability is Japanese HAL. HAL uses EMGs to identify people's sports consciousness. The project considers human legs' viscous and elastic characteristics, researches the HAL's control of viscous characteristics based on impedance control method, and analyzes muscle's viscoelastic characteristics deeply, making HAL's operator feel very comfortable when moving [53–60].

There are both advantages and disadvantages when using myoelectricity sensors. First of all, the central nervous system of the human brain sends out a signal to change the human surface muscle electrical signal, which precedes the muscle's contraction or flexion and extension, so the precedence can make the controller have plenty of time to process electromyograms and compute and control the output signal, offsetting the delay existed in the control system. The above is the favorable aspect. Secondly, myoelectricity sensor has its inherent shortcomings: it is impossible to find a one-to-one relationship between joint torque and muscle's electromyograms; in order to determine the relationship between muscle force and joint torque, the arm of force needs to be determined, and usually the arm of force changes with joint angle. Because the relationship between the arm of force and the EMG signals, the relationship between the arm of force and the muscle forces vary with different individuals, so any controller based on EMG signal is a personalized equipment for a certain operator [108]; under the strenuous exercise, the sensor is easy to fall off and shift; after moving for a long time, the human body' sweating affects the measurement of the sensor; EMG signal often contains a strong noise, which must undergo additional handling before being used in the system; the sensor must be sticked to the body surface whenever used, which is inconvenient.

1.3.4 Master-Slave Control

Master-slave control is generally used for remote robot operating system, and the purpose is to make the distant machine simulate the operator's action. In order to continuously capture the movement of the body, the operator must wear a set of master exoskeleton device. The target of this method is to get human body's joint angle information with the measurement device on the master exoskeleton suit, thus serving as the slave exoskeleton suit's reference input signal to control the slave exoskeleton suit's joint angles and track the corresponding joint angles of the human body. This control method is not only to control the position and direction of the exoskeleton terminal but also to control the pose of the whole exoskeleton suit. Hardiman uses the master-slave control method, whose internal exoskeleton is the "master," controlled by the operator, sending command to external "slave" exoskeleton [21]. The exoskeleton suit of Nanyang Technological University has also adopted the master-slave control method used to control the swinging leg's tracking the human body movement [73]. In the master-slave control method, human must be able to move from the initial position to a desired position to drive the exoskeletons movement, and therefore, when the system is designed, the human and main exoskeleton's activity space must be prepared in the slave exoskeleton in advance, making the design of the system quite complicated.

1.3.5 Force Feedback Control

The feedback information of force sensors can be used in robot force control system to keep the force between the machine and its surrounding environment within a predefined level. Eppinger and Whitney have described the application of a traditional force feedback control in the mechanical arm control [109]. In the exoskeleton system, the force between the operator and the exoskeleton also can be controlled, which will make the operator not feel the presence of exoskeleton. Kazerooni and Hayashibara have adopted the force feedback control method, respectively, to control the upper limb power assistance arm used to enhance people's ability of lifting weights [27, 28, 110]. There are two groups of force sensors in the system, respectively, used to measure the acting force between the human and the machine and between the environment (load) and the machine. The control target is not to make the force on the human and the machine's contact point reach a predefined value, but make the human's force on the load decrease proportionally. In this control method, the machine is used to bear load, people will feel load which decreases proportionally.

In theory, in the force feedback control method, all contact points of human and the machine must be measured with the force sensors, but in fact, it is difficult to realize; therefore, the number and location of the force sensors to be installed must be determined carefully. For example, the force sensors are not suitable for installing in the thigh and shank, because it is difficult for human's thighs and shanks to endure long-time friction, and these places are not only roughly, but also having no rigid interface. The number of force sensors must be greater than or equal to the maximum degrees of freedom of exoskeleton. The appropriate sensor installation position should be on which people are used to feeling the contact force, such as the soles of the feet, or on which the force distributes easily, such as the back [21]. This book will make a thorough research in the force control method of exoskeleton suit theoretically.

1.3.6 Ground Reaction Force Control

Generalized ground reaction force (GRF) control is a kind of control strategy different from the traditional robot control method (the generalized force acting on a point containing the force acting on this point and the torque acting on the segment through this point). From the intuitive, in the process of gait, in addition to gravity, GRF is the only external force and torque acting on human body, and which is the only force to drive the system centroid to move. This allows us to think about whether we can control the exoskeleton by controlling GRF. Because the quality characteristics of exoskeleton are similar to that of human, so if the ground reaction force of the exoskeleton is similar to that of human, the exoskeleton can walk synchronizing with human. The human ground reaction force and exoskeleton ground reaction force must be measured before applying GRF control law. Moreover, all the system's movement characteristics must be measured, and the dynamic model of the exoskeleton and human must be known accurately. Although getting the dynamic model of the exoskeleton is not difficult, to estimate the quality characteristics of human are very difficult, Because the quality characteristics is different for different persons, so we must re-correct the controller according to the different dynamic data of different operators. Furthermore, the geometric features of all exoskeleton's segments must agree with that of human; in order to make the control law work normally, exoskeleton must be customized according to the operator. If the controller is designed when the geometric features of the exoskeleton and human are different, the system equation will become quite complex and requires a lot of anthropometric data [21].

1.3.7 ZMP Control

Xiaopeng Liu and the others of Nanyang Technological University adopt a lower limb trajectory tracking and zero-moment point (ZMP) control methods to control the lower limb power assistance exoskeleton [70, 72, 73]. Using the angle sensor installed on the leg measures human body's position signal, to control exoskeleton's legs' tracing the motion trail of human legs and then to maintain the stability of exoskeleton by controlling the exoskeleton ZMP. The ZMP is defined as a certain point on ground, on which the horizontal component of torque caused by human (or exoskeleton)'s all inertial force and gravity is zero. In the process of walking, only the ZMP trajectory is kept in support area, can the gait's stability be maintained. This method also uses another concept "ground contact point" (GCP): it is defined as a certain point on foot, on which both the ground reaction force and reaction torque are vertical to the ground. In a complete gait, when in the single support phase, support foot's GCP is the exoskeleton's ZMP, and when in the double support phase, the ZMP's position is the function of two supporting feet's GCP position. In the process of control, measuring the ZMP, use the value as the reference input of exoskeleton ZMP to control the exoskeleton ZMP's tracking human ZMP. Once the exoskeleton ZMP is deviated of human ZMP, use actuator on the hip joint to control the exoskeleton's trunk to make compensation for the exoskeleton ZMP, maintaining the ZMP is agreeing with human ZMP. The shortcoming of this method is that the angle sensor must be installed on the human body as the location reference input of exoskeleton, which is not convenient.

1.3.8 Sensitivity Amplification Control

As a human-machine interaction system, the exoskeleton suit needs to complete the human-machine information transferring most comfortably. If complex sensors are installed on the human body, it is not only inconvenient to use but also the human body's movement will affect the sensors' installation position, which will affect the measurement result and will reduce people's comfort. Therefore, to find a natural and direct-interaction method is very important. Berkeley's BLEEX exoskeleton suit has adopted a control method called sensitivity amplification control (SAC), which does not need to install any sensor between human and the machine when controlling the exoskeleton suit following the operator's movement [17, 21, 22, 24, 34]. In this method, the transfer function from the exoskeleton output angle to the force that the human acts on the exoskeleton suit is defined as the sensitivity function. The control goal is to maximize the sensitivity function through the design of the controller, and then, the location of the exoskeleton can be changed with a very small force. However, sensitivity control method is heavily dependent on the dynamic model of the system, and exoskeleton suit is a multi-rigid body, multiple degrees of freedom nonlinear system; therefore, a lot of effort must be taken to establish a more accurate mathematical model of exoskeleton suit [26, 111].

1.4 Major Contents of the Book

The research contents in exoskeleton suit are very rich, including the biomechanics, human-machine coupling, control methods, driving mechanism design, sensors, energy, and communications. The contents are so extensive that one will do not know how to start. Even just the control methods are so various, and there are countless links among them. Therefore, the book mainly chooses some technologies to introduce such as sensitivity amplification control method, and the theory and method of force feedback.

Chapter 1, the Introduction, introduces the research background of this book, exoskeleton suit system's research status at home and abroad and analyzes the development status of exoskeleton suit control methods.

Chapter 2 is this book's research foundation, analyzing the mathematical model of study object from the two aspects of kinematics and dynamics. According to exoskeleton suit's mechanical models, for simplifying the analysis of exoskeleton suit's movement, dividing each leg into supporting model and swing model, from the two aspects of kinematics and dynamics, analyzes the kinematics and dynamics of these two models and provides modeling process, modeling methods, the kinematics, and dynamics model.

Chapter 3 introduces the basic thought and theoretical basis of sensitivity amplification control method in detail. Using classic human-machine interaction model realizes the exoskeleton suit sensitivity amplification control simulation of human-machine integration and provides the comparison result in the computer. Also introduces an exoskeleton suit sensitivity amplification control method based on neural networks, proves its stability, and provides the simulation results.

From the Chaps. 4–7, the book first analyzes the similarities and differences between the exoskeleton suit force and robot force control methods and then introduces the four exoskeleton suit force control methods based on multi-dimensional force/torque sensors. As followings:

Chapter 4 introduces the direct force control method of exoskeleton suit. In this method, after the feedback force is transformed by Jacobian, using the static model compensation method, the control torque of joint space is directly generated, to control the exoskeleton suit's movement, and ensure the acting force between the human and the machine is small. In other words, the control force that human acts on exoskeleton suit is reduced, thereby reducing the energy consumption of the body.

Chapter 5 introduces the exoskeleton suit's force control method based on the inner position loop. For this method, first the exoskeleton suit's position control loop is designed, ensuring that the exoskeleton suit can move accurately according to the reference trajectory. Then, with the feedback information of multi-dimensional force/torque transducer force, the reference trajectory of inner position loop in operating space is generated so as to control exoskeleton suit following the human's movement. And also two position inner loop controllers based on the static model and dynamic model are introduced.

Chapter 6 introduces the exoskeleton suit's impedance control method. Robots' impedance control usually uses the force information to estimate the correction of reference trajectory, and the system's reference trajectory, reference speed, and reference acceleration are known. Exoskeleton suit's impedance control is different from that of common robots, whose reference trajectory, reference speed, and reference acceleration are unknown; thus, the reference trajectory, reference speed, and reference acceleration are estimated at the same time. Because the force feedback information contains noise signal, the differential method will cause the reference speed and the reference acceleration signal are not available. Therefore, the book introduces an exoskeleton suit impedance control method based on tracking differentiator, which can solve estimation problems of reference speed and reference acceleration under the condition of large noise and achieve the zero force compliance control.

Chapter 7 introduces the impedance control method of exoskeleton suit with uncertainty. Exoskeleton suit's impedance controller is designed based on a standard model, while the actual exoskeleton suit's load need often change, and the change range is large. So the controller designed on the standard model cannot meet the performance requirements of load changes, when the acting force between human and the machine will increase, thus cause the human body's energy expenditure increase. The impedance control method of exoskeleton suit with uncertainty introduced in this book uses RBF neural networks to estimate the system's uncertainty, and compensate in the control algorithm to reduce the influence of system's uncertainty on the human-machine interaction force.

Chapter 8 introduces exoskeleton suit reference trajectory estimation method based on RBF neural networks. There are two kinds of reference trajectory estimation method. One is based on the model estimation, which uses the fixed human-machine interaction model to estimate the reference trajectory of the body, because the human-machine interaction model changes with the different environment and individuals, so the method is not universal. The other is based on the neural networks, because the neural networks can realize the nonlinear mapping from input to output and has the function of real-time online learning, so it can achieve the reference trajectory estimation under the condition of different environment and different human bodies.

Chapter 9 summarizes the main contents of the book, makes comparative analysis on the control effect of all kinds of force control methods introduced in this book, and makes an outlook on the further research direction of the force control.

References

1. Yang Zhiyong, Gui Lihua, Yang Xiuxia, Gu Wenjin. On Virtual Force Control of Exoskeleton Suit [J]. Robot, 2009, 31 (5).
2. Yang Zhiyong, Gui Lihua, Yang Xiuxia, Gu Wenjin. Exoskeleton Suit Neural Network Sensitivity Amplification Technique [J]. Journal of Jilin University Engineering and Technology Edition, 2009, 39 (3): 824-829.
3. Yang Zhiyong. Exoskeleton Suit's Force Control Method Research and the Prototype Design for Individual Load [D]. Naval Aeronautical Engineering Institute, Doctor, Yantai: 2009.
4. Chen Ying, Yang Canjun. The Human-Machine Intelligent System [M]. Hangzhou: Zhejiang University Press, 2006.
5. Peter, Neuhaus, Kazerooni H. Design and Control of Human Assisted Walking Robot [C]. Proceedings of 2000 IEEE International Conference on Robotics & Automation, San Francisco, 2000: 563–569.
6. Vukobratovic, M, Ciric V, Hristic D. Contribution to the study of active exoskeletons [C]. Proceedings of the 5th International Federation of Automatic Control Congress, Paris, France, 1972: 13–19.

7. Hristic, D, M Vukobratovic. Development of Active Aids for Handicapped [C]. Proceedings of the III International Conference on Biomedical Engineering, Sorrento, Italy, 1973: 123–129.
8. Aaron, M. Dollar, Hugh Herr. Lower Extremity Exoskeletons and Active Orthoses: Challenges and State-of-the-Art [J]. IEEE Transactions on Robotics, 2008, 24 (1): 144–158.
9. Colombo, Gery, Matthias Jorg, Volker Dietz. Driven Gait Orthosis to do Locomotor Training of Paraplegic Patients [C]. Proceedings of the 22nd Annual EMBS International Conference, Chicago IL, 2000: 3159–3163.
10. Daniel, P. Ferris, Gregory S. Sawicki, Antoinette R. Domingo. Powered Lower Limb Orthoses for Gait Rehabilitation [J]. Topics in Spinal Cord Injury Rehabilitation, 2005, 11 (2): 34–49.
11. Ekkelenkamp, R., Veneman Jan F., H. Van Der Kooij. LOPES: Selective control of gait functions during the gait rehabilitation of CVA patients [C]. Proceedings of the 2005 IEEE 9th International Conference on Rehabilitation Robotics, Chicago, IL, USA, 2005: 361–364.
12. Nelson, Costa, Darwin G. Caldwell. Control of a Biomimetic "Soft-actuated" 10 DoF Lower Body Exoskeleton [J].
13. Pieter, Beyl, Joris Naudet, Ronald Van Ham, Dirk Lefeber. Mechanical Design of an Active Knee Orthosis for Gait Rehabilitation [C]. Proceedings of the 2007 IEEE 10th International Conference on Rehabilitation Robotics, Noordwijk, The Netherlands, 2007: 100–105.
14. Veneman, J F, R. Ekkelenkamp, R. Kruidhof, F. C. T. Van Der Helm, H. Van Der Kooij. Design of a series elastic- and bowdencale-based actuation system for use as torque-actuator in exoskeleton-type training [C]. Proceedings of the 9th International Conference on Rehabilitation Robotics, Chicago, IL, USA, 2005: 496–499.
15. Veneman, Jan F., Rik Kruidhof, Edsko E. G. Hekman, Ralf Ekkelenkamp, Edwin H. F. Van Asseldonk, Herman Van Der Kooij. Design and Evaluation of the LOPES Exoskeleton Robot for Interactive Gait Rehabilitation [J]. IEEE Transactions on Neural Systems and Rehabilitation Engineering, 2007, 15 (3): 379–386.
16. "Hardiman." http://www.davidszondy.com/future/robot/hardiman.htm.
17. Kazerooni, H, Steger R. The Berkeley Lower Extremity Exoskeleton [J]. Journal of Dynamic Systems, Measurement, and Control, 2005, 128 (3): 14–25.
18. "Exoskeletons for Human Performance Augmentation (EHPA)." http://www.beachbrowser.com/Archives/Science-and-Health/January-2001/Exoskeletons-for-Human-Augmentation.htm.
19. Richardson, B. S., J. F. Jansen, J. F. Birdwell, A. C. Boynton, Iii H. P. Crowell, W. K. Durfee, J. D. Gongola, S. M. Killough, D. J. Leo, R. F. Lind, L. J. Love, M. Mungiole, F. G. Pin, J. C. Rowe, O. A. Velev, T. Zambrano. Phase I Report: DARPA Exoskeleton Program [R]. ORNL/TM-2003/216, 2004.
20. John, Jansen, Brad Richardson, Francois Pin, Randy Lind, Joe Birdwell. Exoskeleton for Soldier Enhancement Systems Feasibility Study [R]. ORNL/TM-2000/256, 2000.
21. Racine, Jean-Louis Charles. Control of a Lower Extremity Exoskeleton for Human Performance Amplification [D]. University of California, Ph. D, Berkeley: 2003.
22. Kazerooni, H, Huang L H, Steger R. On the Control of the Berkeley lower extremity exoskeleton (BLEEX) [C]. Proceedings of IEEE International Conference on Robotics and Automation (ICRA 2005), Barcelona, Spain, 2005: 4353–4360.
23. Chu, Andrew, Kazerooni H, Adam B Zoss. On the Biomechanical Design of the Berkeley Lower Extremity Exoskeleton (BLEEX) [C]. Proceedings of the 2005 IEEE International Conference on Robotics and Automation, Barcelona, Spain, 2005: 4345–4352.
24. Huang, Lihua. Robotics Locomotion Control [D]. University of California, Ph. D, Berkeley: 2005.
25. Zoss, Adam B., Kazerooni H, Andrew Chu. Biomechanical design of the Berkeley lower extremity exoskeleton (BLEEX) [J]. IEEE/ASME Transactions on Mechatronics, 2006, 11 (2): 128–138.
26. Ghan, Justin, Ryan Steger, Kazerooni H. Control and System Identification for the Berkeley Lower Extremity Exoskeleton(BLEEX) [J]. Advanced Robotics, 2006, 20 (9): 989–1014.

27. Kazerooni, H. Human/ Robot Interaction via the Transfer of Power and Information Signals, Part I: Dynamics and Control Analysis [C]. IEEE International Conference on Robotics and Automation, 1989: 1632–1640.
28. Kazerooni, H. Human/ Robot Interaction via the Transfer of Power and Information Signals, Part II: An Experimental Analysis [C]. IEEE International Conference on Robotics and Automation, 1989: 1641–1647.
29. Zoss, Adam, Kazerooni H. Design of an electrically actuated lower extremity exoskeleton [J]. Advanced Robotics, 2006, 20 (9): 967–988.
30. Amundson, Kurt, Justin Raade, Nathan Harding, Kazerooni H. Development of hybrid hydraulic–electric power units for field and service robots [J]. Advanced Robotics, 2006, 20 (9): 1015–1034.
31. Chu, Andrew. Design of the Berkeley Lower Extremity Exoskeleton (BLEEX) [D]. University of California, Ph. D., Berkeley: 2005.
32. Steger, John Ryan. A design and control methodology for human exoskeletons [D]. University of California, Ph.D, Berkeley: 2006.
33. Zoss, Adam Brain. Actuation Design and Implementation for Lower Extremity Human Exoskeletons [D]. University of California, Ph.D, Berkeley: 2006.
34. Ryan, Steger, Sung Hoon Kim, Kazerooni H. Control Scheme and Networked Control Architecture for the Berkeley Lower Extremity Exoskeleton (BLEEX) [C]. Proceedings of the 2006 International Conference on Robotics and Automation, Orlando, Florida, 2006: 3469–3477.
35. Chu, Andrew. Design of the Berkeley Lower Extremity Exoskeleton (BLEEX) [D]. University of California, Ph. D, Berkeley: 2005.
36. Meng Zhaoyu. Rescue Armor [J]. Popular Science, 2008, (6): 64–75.
37. Conor, James Walsh, Ken Endo, Hugh Herr. A Quasi-Passive Leg Exoskeleton for Load-Carring Augmentation [J]. International Journal of Humanoid Robotics, 2007, (5): 1 ~ 21.
38. Hugh, Herr, Ari Wilkenfeld. User-adaptive Control of a Magneto rheological prosthetic knee [J]. Industrial Robot: An International Journal, 2003, 30 (1): 42–55.
39. Conor, James Walsh, Kenneth Pasch Scd Pe, Herr Hugh. An autonomous, under actuated exoskeleton for load-carrying augmentation [C]. Proceedings of the 2006 IEEE/RSJ International Conference on Intelligent Robotics and Systems, Beijing, China, 2006: 1410–1415.
40. Conor, Jame Walsh, Daniel Paluska, Kenneth Pasch, William Grand, Andrew Valiente, Herr Hugh. Development of a lightweight under actuated exoskeleton for load-carrying augmentation [C]. Proceedings of the 2006 IEEE International Conference on Robotics and Automation, Orlando, Florida, 2006: 3485–3491.
41. Wang Yong, Yang Jie. The Status of Study Based on Passive-Dynamic Bipedal Robot [J]. Robot Technique and Application, 2005, (6): 31–33.
42. Gong Mingming, Zhu Weibing, Li Feng. Key Research Point of Passive Dynamic Walking [J]. Mechanical Engineer, 2006, (8): 106–108.
43. Steve, Collins, Andy Ruina, Russ Tedrake, Martijn Wisse. Efficient Bipedal Robots Based on Passive-Dynamic Walkers [J]. Science, 2005, 307 (18): 1082–1085.
44. Timothy, G. Mcgee, Justin W. Raade, Homayoon Kazerooni. Theoretical Analysis and Experimental Verification of a Monopropellant Driven Free Piston Hydraulic Pump [C]. Proceedings of 2003 ASME International Mechanical Engineering Congress & Exposition, Washington, D. C., 2003: 1–7.
45. Justin, William Raade. Graphical Analysis of Power Systems for Mobile Robotics [D]. University of California, Ph. D, Berkeley: 2006.
46. Sunghoon, Kim, George Anwar, Kazerooni H. High-speed Communication Network for Controls with the Application on the Exoskeleton [C]. Proceedings of the 2004 American Control Conference, Boston, Massachusetts, 2004: 355–360.
47. Sunghoon, Kim. Design and Analysis of High-speed Ring-based Networked Control Systems for Real-time Application [D]. University of California, Ph. D, Berkeley: 2005.

48. Jason, Wheeler, Brandon Rohrer, Deepesh Kholwadwala, Stephen Buerger, Richard Givler, Jason Neely, Clint Hobart, Paul Galambos. In-Sole MEMS Pressure Sensing for a Lower-Extremity Exoskeleton [C]. The First IEEE/RAS-EMBS International Conference on Biomedical Robotics and Biomechatronics, Pisa, Italy, 2006: 31–34.
49. Joseph, J. Misuraca, Constantinos Mavroidis. Lower Limb Human Muscle Enhancer [C]. Proceedings of IMECE01: International Mechanical Engineering Conference and Exposition, New York, New York, 2001: 1–7.
50. "berkeley bionics." http://www.berkeleybionics.com/Companyprofile.html.
51. Meng Ma. Freedom of Walking [J]. 21st Century Business Review, 2010, (11): 100–101.
52. "XOS-2." http://www.workercn.cn/.
53. Okamura, J, Tanaka H, Sankai Y. EMG-based Prototype Powered Assistive system for Walking Aid [C]. Proceedings of Asian Symposium on Industrial Automation and Robotics (ASIAR' 99), Bangkok, Thailand, 1999: 229–234.
54. Lee, S, Sankai Y. Power Assist Control for Walking Aid with HAL-3 Based on EMG and Impedance Adjustment around Knee Joint [C]. Proceedings of IEEE/RSJ International Conf on Intelligent Robots and Systems (IROS 2002), EPFL, Switzerland, 2002: 1499–1504.
55. Lee, S, Sankai Y. Power assist control for leg with HAL-3 based on virtual torque and impedance adjustment [C]. Proceedings of IEEE International Conference on Systems, Man and Cybernetics (SMC), Hammamet, Tunisia, 2002: TP1B3 (CD-ROM).
56. Kawamoto, Hiroaki, Yoshiyuki Sankai. Power assist system HAL-3 for gait disorder person [C]. International Conference on Computers Helping People with Special Needs, Linz, Austria, 2002: 196–203.
57. Kawamoto, Hiroaki, Yoshiyuki Sankai. Comfortable Power Assist Control Method for Walking Aid by HAL-3 [C]. Proceedings of IEEE International Conference on System, Man, and Cybernetics (SMC 2002), Hammamet, Tunisia, 2002: 190–193.
58. Kawamoto, Hiroaki, Yoshiyuki Sankai. Power Assist Method Based on Phase Sequence Driven by Interaction between Human and Robot Suit [C]. Proceedings of the 2004 IEEE International Workshop on Robot and Human Interactive Communication, Kurashiki, Okayama Japan, 2004: 491–496.
59. Kawamoto, Hiroaki, Shigehiro Kanbe, Yoshiyuki Sankai. Power Assist Method for HAL-3 Estimating Operator's Insertion Based on Motion Information [C]. Proceedings of the 2003 IEEE International Workshop on Robot and Human Interactive Communication, Millbrae, California, USA, 2003: 67–72.
60. Kota, Kasaoka, Yoshiyuki Sankai. Predictive Control Estimating Operator's Intention for Stepping-up Motion by Exoskeleton Type Power Assist System HAL [C]. Proceedings of 2001 IEEE/RSJ International Conference on Intelligent Robots and Systems, Maui, Hawaii, USA, 2001: 1578–1583.
61. "sanlab HAL." http://sanlab.kz.tsukuba.ac.jp/HAL/indexE.html.
62. Kawamoto, Hiroaki, Tomoya Shiraki, Tasuku Otsuka, Yoshiyuki Sankai. Meal-Assistance by Robot Suit HAL using Detection of Food Position with Camera [C]. Proceedings of the 2011 IEEE International Conference on Robotics and Biomimetics, Phuket Thailand, 2011: 889–894.
63. Yamamoto, K., M. Ishii. Stand Alone Wearable Power Assisting Suit—Sensing and Control Systems [C]. 13th IEEE International Workshop on Robot and Human Interactive Communication, Roma, Italy, 2004: 661–666.
64. Yoshiitsu, T., K. Yamamoto. Development of a Power Assist Suit for Nursing Work [C]. SCIE 2004 Annual Conference, Sapporo, Tokyo, 2004: 577–580.
65. Kim, Yoon Sang, Jangwook Lee, Sooyong Lee, Munsang Kim. A Force Reflected Exoskeleton-Type Master arm for Human-Robot Interactions [J]. IEEE Transactions on System, Man, and Cybernetics-Part A: Systems and Humans, 2005, 35 (2): 198–212.
66. Kim, Wan-Soo, Seung-Hoon Lee, Hee-Don Lee, Seung-Nam Yu, Jung-Soo Han, Chang-Soo Han. Development of the Heavy Load Transferring Task Oriented Exoskeleton Adapted by Lower Extremity Using Qausi-active Joints [C]. ICROS-SICE International Joint Conference, Fukuoka, Japan, 2009: 1353–1358.

67. Kong, Kyoungchul, Joonbum Bae, Masayoshi Tomizuka. Control of Rotary Series Elastic Actuator for Ideal Force-Mode Actuation in Human-Robot Interaction Applications [J]. IEEE/ASME Transactions on Mechatronics, 2009, 14 (1): 105–118.
68. Kong, Kyoungchul, Masayoshi Tomizuka. A Gait Monitoring System Based on Air Pressure Sensors Embedded in a Shoe [J]. IEEE/ASME Transactions on Mechatronics, 2009, 14 (3): 358–370.
69. Kong, Kyoungchul, Masayoshi Tomizuka. Control of Exoskeletons Inspired by Fictitious Gain in Human Model [J]. IEEE/ASME Transactions on Mechatronics, 2009, 14 (6): 689–698.
70. Liu, Xiaopeng, K. H. Low. Development and Preliminary Study of the NTU Lower Extremity Exoskeleton [C]. Proceedings of the 2004 IEEE Conference on Cybernetics and Intelligent Systems, Singapore, 2004: 1243–1247.
71. Liu, Xiaopeng, K. H. Low, Hao Yong Yu. Development of a Lower Extremity Exoskeleton for Human Performance Enhancement [C]. Proceedings of the 2004 IEEE/RSJ International Conference on Intelligent Robots and Systems, Sendai, Japan, 2004: 3889–3894.
72. Low, K. H., Xiaopeng Liu, C. H. Goh, Hao Yong Yu. Locomotive Control of a Wearable Lower Exoskeleton for Walking Enhancement [J]. Journal of Vibration and Control, 2006, 12 (12): 1311–1316.
73. Low, K. H., Xiaopeng Liu, Hao Yong Yu, Hendra S Kasim. Development of a Lower Extremity Exoskeleton—Preliminary Study for Dynamic Walking [C]. The 8th International Conference on Control Automation, Robotics and Vision, Kunming, China, 2004: 2088–2093.
74. Jerry, E. Pratt, Benjamin T. Krupp, Christopher J. Morse, Steven H. Collins. The RoboKnee: An Exoskeleton for Enhancing Strength and Endurance During Walking [C]. Proceedings of the 2004 IEEE International Conference on Robotics & Automation, New Orleans, LA, 2004: 2430–2435.
75. Bergamasco, Massimo, Fabio Salsedo, Simone Marcheschi, Nicola Lucchesi, Marco Fontana. A Novel Compact and Lightweight Actuator for Wearable Robots [C]. IEEE International Conference on Robotics and Automation, Anchorage, Alaska, USA, 2010: 4197–4203.
76. Keitaro, Naruse, Satoshi Kawai, Hiroshi Yokoi, Yukinori Kakazu. Development of Wearable Exoskeleton Power Assist System for Lower Back Support [C]. Proceedings of the 2003 IEEE/RSJ International Conference on Intelligent Robots and Systems, Las Vegas, Nevada, 2003: 3630–3635.
77. Keitaro, Naruse, Satoshi Kawai, Takuji Kukichi. Three-dimensional Lifting-up Motion Analysis for Wearable Power Assist Device of Lower Back Support [C]. IEEE/RSJ International Conference on Intelligent Robots and Systems, 2005: 2959–2964.
78. Rahman, S. M. Mizanoor, Ryojun Ikeura, Soichiro Hayakawa, Hideki Sawai. Design and Control of a Power Assist System for Lifting Objects Based on Human Operator's Weight Perception and Load Force Characteristics [J]. IEEE Transactions on Industrial Electronics, 2011, 58 (8): 3141–3150.
79. Rossi, S.M.M. De, T. Lenzi, N. Vitiello, M. Donati, A. Persichetti, F. Giovacchini, F. Vecchi, M. C. Carrozza. Development of an in-shoe Pressure-Sensitive Device for Gait Analysis [C]. 33rd Annual International Conference of the IEEE EMBS, Boston, Massachusetts USA, 2011: 5637–5640.
80. Gabriel, Aguirre-Ollinger, J. Edward Colgate, Michael A. Peshkin, Ambarish Goswami. A 1-DOF Assistive Exoskeleton with Virtual Negative Damping: Effects on the Kinematic Response of the Lower Limbs [C]. Proceedings of the 2007 IEEE/RSJ International Conference on Intelligent Robots and Systems, San Diego, CA, USA, 2007: 1938–1944.
81. Narari, Amir, Nikos G. Tsagarakis, Bram Vanderborght, Darwin G.Caldwell. A Novel Actuator with Adjustable Stiffness (AwAs) [C]. The 2010 IEEE/RSJ International Conference on Intelligent Robots and Systems, Taipei, Taiwan, 2010: 4201–4206.
82. Aguirre-Ollinger, Gabriel, J. Edward Colgate, Michael A. Peshkin, Ambarrish Goswani. Inertia Compensation Control of a One-Degree-of-Freedom Exoskeleton for Lower-Limb

Assistance: Initial Experiments [J]. IEEE Transitions on Neural Systems and Rehabilitation Engineering, 2012, 20 (1): 68–77.
83. Shunji, Moromugi. Exoskeleton Suit for Human Motion Assistance [D]. University of California, Ph.D, Irvine: 2003.
84. Niu Bin. The Control Mechanism Study and Realization of Wearable Lower Extremity Exoskeleton [D]. Zhejiang University, Master's Thesis, Hangzhou: 2006.
85. Liu Zhijuan. The Control System Research of Multi-Degree of Freedom Lower Extremity Exoskeleton [D]. Zhejiang University, Master's Thesis, Hangzhou: 2011.
86. Chen Feng. The Technical Research of Wearable Walking Power Assist Robot [D]. University of Science and Technology of China, Doctoral Dissertation, Hefei: 2007.
87. Yao Junzhang. The Control System Research of Power-assisted robot's Frequency Multiplication Algorithm and Hip Joint's Parallel Mechanism [D]. University of Science and Technology of China, Master's Thesis, Hefei: 2011.
88. Heng, Cao, Wenjin Gu, Yuhai Yin, Zhiyong Yang. Neural-Network Inverse Dynamic Online Learning Control on Physical Exoskeleton [C]. 13th International Conference of Neural Information Processing, ICONIP 2006, Hong Kong, 2006: 702–710.
89. Heng, Cao, Yuhai Yin, Zhengyang Ling. Walk-aided System with Wearable Lower Extremity Exoskeleton for Brain-machine Engineering [C]. Proceedings of the International Conference on Cognitive Neurodynamic, shanghai, China, 2007: 849–855.
90. Heng, Wang, K. H. Low, Michael Yu Wang. Reference Trajectory Generation for Force Tracking Impedance Control by Using Neural Network-based Environment Estimation [C]. 2006 IEEE Conference on Robotics Automation and Mechatronics, Bangkok, Thailand, 2006: 1–6.
91. Lei Bing. The Structure Optimization and Performance Evaluation Research of Power-assisted Walking Mechanical Legs [D]. East China University Of Science and Technology, Master's Thesis, Shanghai: 2011.
92. Li Xiaoming. The Robot's Remote Control Based on the Exoskeleton Technology [D]. Zhejiang University, Doctoral Dissertation, Hangzhou: 2004.
93. Wang Lan, Wang Ting, Zhang Lixun, Deng Zongquan. The Simulation Research of Power-assisted Mechanical Leg [J]. Journal of Machine Design, 2006, 23 (9): 12–15.
94. Lan, Wang, Zhongquan Deng, Lixun Zhang, Qingxin Meng. Analysis of Assistant Robotic Leg on MATLAB [C]. Proceedings of 2006 IEEE International Conference on Mechatronics and Automation, Luoyang, China, 2006: 1092–1096.
95. Deng, Xiaohong, Huanghuan Shen, Feng Chen, Yong Yu, Yunjian Ge. Motion Information Acquisition from Human lower Limbs for wearable Robot [C]. Proceedings of the 2007 International Conference on Information Acquisition, Jeju City, Korea, 2007: 137–142.
96. Feng, Chen, Yong Yu, Yunjan Ge. Dynamic Model and Motion Control Analysis of the Power Assist Intelligence Leg [C]. Proceedings of the 6th World Congress on Intelligent Control and Automation, Dalian, China, 2006: 6346–6350.
97. Zhao Yanjun. The Working Mechanism Research of Human Body's Lower Extremity Exoskeleton [D]. Nanjing University of Science and Technology, Master's Thesis, Nanjing: 2006.
98. Han, Yali, Wang Xingsong. Kinematics Analysis of Lower Extremity Exoskeleton [C]. 2008 Chinese Control and Decision Conference, Yantai, China, 2008: 2753–2758.
99. Chen, Jinzhou, Wei-Hsin Liao. Design and Control of a Magneto rheological Actuator for Leg Exoskeleton [C]. Proceedings of the 2007 IEEE International Conference on Robotics and Biomimetic, Sanya, China, 2007: 1388–1393.
100. Yano, H., Kaneko S., Nakazawa K., Yamamoto S-L, Bettoh A. A New Concept of Dynamic Orthosis for Paraplegia: the Weight Bearing Control (WBC) Orthosis [J]. Prosthetics and Orthotics International, 1997, 21 (3): 222–228.
101. Gui Lihua, Yang Zhiyong, Gu Wenjin, Zhang Yuanshan, Yang Xiuxia. Development of Power Assistance Exoskeleton Suit (NAEIES) [J]. Journal of Naval Aeronautical Engineering Institute, 2007, 22 (4): 467–470.

102. Ruthenberg, B J, Wasylewski N A, Beard J E. An Experimental Device for Investigating the Force and Power Requirements of a Powered Gait Orthosis [J]. Journal of Rehabilitation Research and Development, 1997, 34 (2): 203–213.
103. Rosen, J., Brand M., Fuchs M.B., Arcan M. A My signal-Based Powered Exoskeleton System [J]. IEEE Transaction on Systems, Man, and Cybernetics Part A: Systems and Humans, 2001, 31 (3): 210–222.
104. Marcello, Mulas, Michele Folgheraiter, Giuseppina Gini. An EMG-controlled Exoskeleton for Hand Rehabilitation [C]. Proceedings of the 2005 IEEE 9th International Conference on Rehabilitation Robotics, Chicago, IL, USA, 2005: 371–374.
105. Matthew, Dicicco, Lenny Lucas, Yoky Matsuoka. Comparison of Control Strategies for an EMG Controlled Orthotic Exoskeleton for the Hand [C]. Proceedings of the 2004 IEEE International Conference on Robotics and Automation, New Orleans, LA, USA, 2004: 1622–1627.
106. H., Hel, K. Kiguchil. A Study on EMG-Based Control of Exoskeleton Robots for Human Lower-limb Motion Assist [C]. 6th International Special Topic Conference on Information Technology Application in Biomedicine (ITAB), Tokyo, Japan, 2007: 292–295.
107. Christian, Fleischer, Christian Reinicke, Gunter Hommel. Predicting the Intended Motion with EMG Signals for an Exoskeleton Orthosis Controller [C]. 2005 IEEE/RSJ International Conference on Intelligent Robots and Systems, Alberta, Canada, 2005: 2029–2034.
108. Kiguchi, K., Iwami K., Saza T., Kariya S., Watanabe K., Izumi K., Fukuda T. A Study of an Exoskeletal Robot for Human Shoulder Motion Support [C]. Proceedings of the 2001 IEEE/RSJ International Conference on Intelligent Robots and systems, Maui Hawaii, USA, 2001: 2111–2116.
109. Eppinger, S.D., Seering W.P. Understanding Bandwidth Limitations in Robot Force Control [C]. Proceeding of the IEEE International Conference on Robotics and Automation, 1987: 904–909.
110. Hayashibara, Y., Tanie K., Arai H. Design of a Power Assist System with Consideration of Actuator's Maximum Torque [C]. Proceedings of the 4th IEEE International Workshop on Robot and Human Communication, New York, USA, 1995: 379–384.
111. Justin, Ghan, H. Kazerooni. System Identification for the Berkeley Lower Extremity Exoskeleton (BLEEX) [C]. Proceedings of the 2006 IEEE International Conference on Robotics and Automation, Orlando, Florida, USA, 2006: 3477–3484.

Chapter 2
Model of Exoskeleton Suit

2.1 Introduction

To facilitate the analysis of the characteristics of the exoskeleton suit system, assume that every link of the exoskeleton suit is rigid connecting rod, and the connecting rods are connected through joints to form a multi-rigid-body chain structure. The abstract model based on the assumption can be used as the object of our theoretical research and analysis.

The mathematical model of multi-connecting rods' object generally consists of two parts: the kinematics and dynamics [1]. In the control of the exoskeleton suit, the position relationship and the speed relationship among the joint connecting rods of exoskeleton suit can be directly measured, namely the angle and the angular velocity of each joint. How to calculate the pose of the end effector in the operating space according to the position and velocity of each connecting rod (the terminal of support phase is gravity center of the body, and the terminal of the swing phase is ankle) is known as exoskeleton suit's kinematic analysis or forward kinematics problem. On the contrary, according to the position of the end effector in the operating space and speed expectation, calculate the corresponding position and speed of each joint's connecting rod reversely, which is called kinematic synthesis or inverse kinematics problem of the robot. And the relationship between exoskeleton suit' movement and the force and torque that produce this movement becomes the dynamics problem, which also can be divided into two problems: according to the torque or force acting on the joints, calculating the joints' position, velocity and acceleration is the forward dynamics problem; according to the desired joints' position, velocity and acceleration, calculating the required torque or force is called inverse dynamics problem [1–5].

This chapter first describes the book's research object, the exoskeleton suit in detail. Then, it analyzes the forward kinematics and inverse dynamics problems based on the mechanical model of an exoskeleton suit and establishes the kinematics and dynamics models of the exoskeleton suit. At the same time, it describes

Z. Yang et al., *Force Control Theory and Method of Human Load Carrying Exoskeleton Suit*, DOI 10.1007/978-3-662-54144-9_2

the human body model, human-machine interaction model, and the virtual prototype research method of the exoskeleton suit and provides the human-machine interaction models that can satisfy the requirement of simulation assumptions.

2.2 Basic Description of Exoskeleton Suit

2.2.1 Mechanical Model of Exoskeleton Suit

From the aspects of energy consumption and realized functions, there are two development trends for the exoskeleton suit.

1. The whole body exoskeleton suit

The whole body-type exoskeleton suit refers to the exoskeleton that contains upper and lower limbs and the exoskeleton whose all main movement joints have actuators. This exoskeleton suit's representative is the "WEAR" developed by SARCOS Company in the USA, which mainly drives such joints as the hip, knee and ankle joints of two legs, and shoulder, elbow, and wrist joints of two arms. The WEAR tries to reduce the operator's energy consumption as much as possible and increase more functions, such as protective armor, the body air-conditioning equipment, color clothes, and flexible fabric (instruments), and even realize the function to fly.

The whole body-type exoskeleton suit looks at the future, in order to meet human's dream of becoming superman. But it requires more drive joints: Not only the structure is complicated, but also energy consumption is large. Now, we only can stay in the laboratory dragging long cable under the condition without miniaturization, high density, and portable energy devices.

2. The lower limb-type exoskeleton suit

Lower limb-type exoskeleton suit refers to the exoskeleton that only contains lower limbs. The representative of the exoskeleton suit is Berkeley's HULC of the USA. The exoskeleton suit's desired function is very simple, that is, to help soldiers carry load and reduce the energy consumption. It drives a few joints selectively, for example, HULC and MIT exoskeleton suit only drives the knee joint, trying to transfer the load to the ground as much as possible.

Because the lower limb-type exoskeleton suit's function is single and the number of its driving joints is reduced, its structure design is relatively simple. And because of its lighter weight, smaller volume, reduced power consumption, and increased portability, the function is easy to be realized under current technical level.

This book mainly introduces the lower limb-type exoskeleton suit, providing an instance of exoskeleton suit as shown in Fig. 2.1 [6]. The exoskeleton suit is composed of seven rigid connecting rods, namely the trunk (back frame), two legs, two shanks, and two feet. The trunk of exoskeleton suit carries the load, and the load can be seen as part of the trunk.

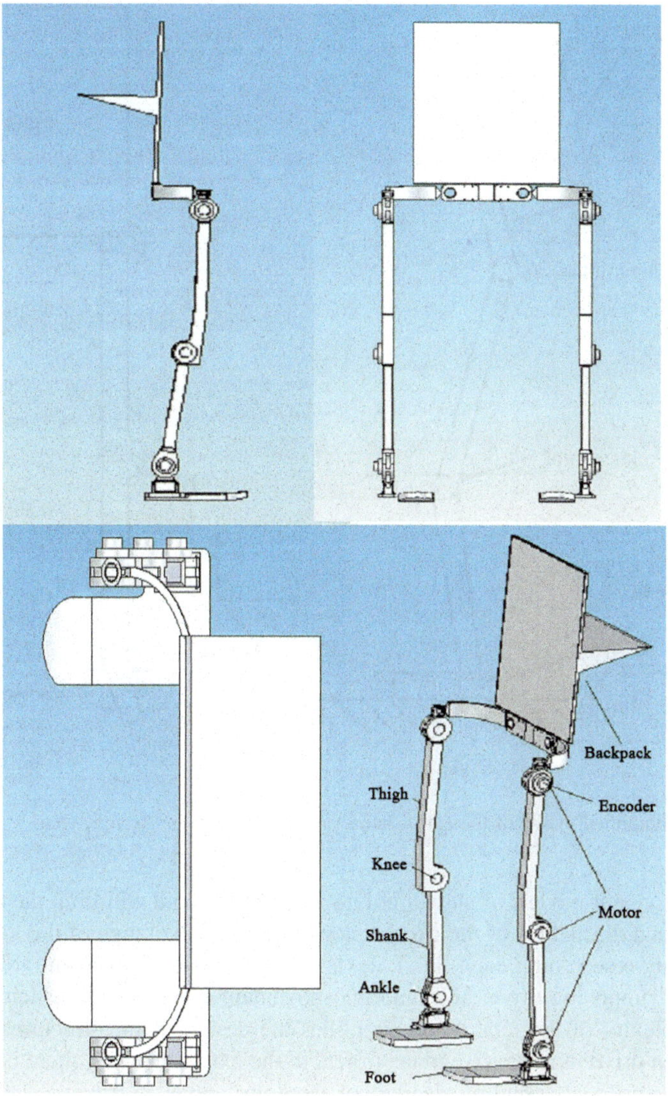

Fig. 2.1 Mechanical model of the exoskeleton suit

Simplify the mechanical model drawings as shown in Fig. 2.1 to plane sketch as shown in Fig. 2.2. The *Y*-axis of the vertical direction and the *X*-axis of the horizontal direction are in a longitudinal plane, as shown in Fig. 2.2a. When analyzing the exoskeleton suit's movement and control theoretically, considering exoskeleton suit's most basic movement form is forward walking in flat ground, and the range of each joint movement in longitudinal plane is bigger than the range in other planes. Therefore, this book will make theoretical analysis and research based on the

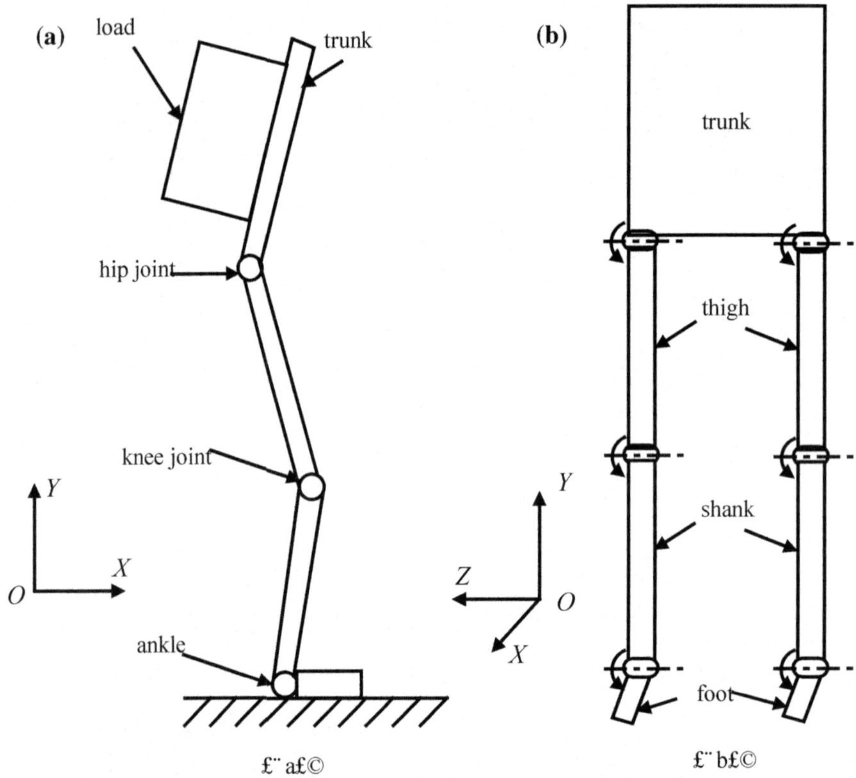

Fig. 2.2 Simplified model of the exoskeleton suit: **a** side view and **b** front view

exoskeleton suit models as shown in Figs. 2.1 and 2.2 and will limit the theoretical analysis and discussion of the exoskeleton suit within the range of the longitudinal plane, only researching each link's flexion and extension movement around coronary axis, imposing drive, and ignoring movement other planes, which decreases the complexity of the model and simplifies the design of the structure. Although there is no drive in transverse plane (refers to the external drive), there is also need to design the corresponding degrees of freedom, and it is driven by the human body's energy, in order to follow the human body's movement; the energy consumption is lesser, so it can be ignored.

2.2.2 Segment Properties of Exoskeleton Suit

In the later parts of this book, we will build the exoskeleton suit's kinematic and dynamic models according to the above exoskeleton suit models. Before modeling, first define the physical properties of the studied exoskeleton suit, as follows: call

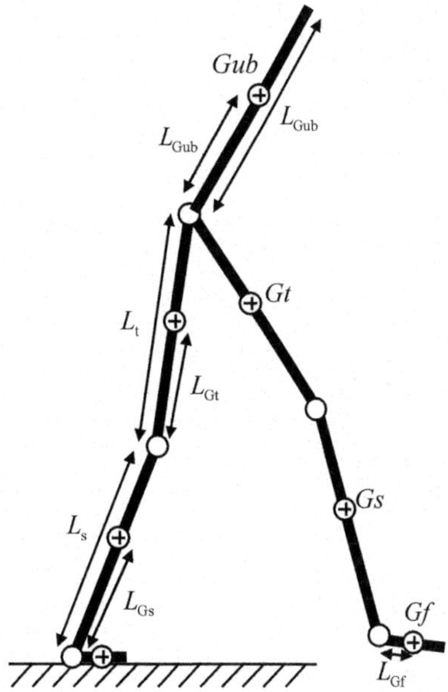

Fig. 2.3 Link properties of the exoskeleton suit

each connecting rod of the exoskeleton suit, namely the thighs, the shanks, feet, and torso, a link. The links' attributes include quality, rotational inertia, and length as shown in Fig. 2.3. These attribute parameters, as the most basic modeling parameter, must be determined in advance.

Define that m represents quality, I represents rotational inertia around the center of mass, G represents the center of mass, L represents the rod length, and L_{G} represents the components from joint points to the center of gravity in the reference coordinate system. Then the parameters of exoskeleton suit's each part are as follows:

(1) Feet: $m_{\mathrm{f}}, I_{\mathrm{f}}, L_{\mathrm{f}}, L_{\mathrm{Gf}}$
(2) Shanks: $m_{\mathrm{s}}, I_{\mathrm{s}}, L_{\mathrm{s}}, L_{\mathrm{Gs}}$
(3) Thighs: $m_{\mathrm{t}}, I_{\mathrm{t}}, L_{\mathrm{t}}, L_{\mathrm{Gt}}$
(4) Trunk: $m_{\mathrm{ub}}, I_{\mathrm{up}}, L_{\mathrm{up}}, L_{\mathrm{Gup}}$.

For the convenience of study, this book ignores the offset of the center of gravity of the exoskeleton suit trunk (including load) in the longitudinal plane. In the theoretical analysis stage of verification, this assumption is reasonable.

2.2.3 Definition of Coordinate System

For both kinematics model and dynamics model, the pose of exoskeleton suit's each connecting rod needs to be defined. To define the pose of each connecting rod, we

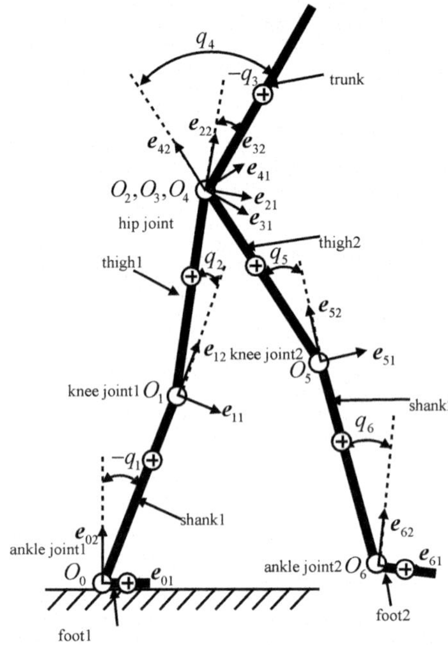

Fig. 2.4 Coordinate system and symbol definition

need first define the local coordinate system of each rod, and using coordinate system describes geometric relationship of all the exoskeleton suit's entities. The local coordinate system defining the exoskeleton suit is shown in Fig. 2.4 [7].

Reference coordinate system (coordinate system 0) is defined in the landing place of the heel, $\boldsymbol{e}_{01}$ is in parallel to the instep, and pointing from ankle to tiptoe, $\boldsymbol{e}_{02}$, is perpendicular to the instep. Other coordinate systems except reference coordinate system are associated with the system status. The other coordinate systems are defined as follows:

(1) Coordinate system 1 is fixed in knee joint 1 of the standing leg 1, and $-\boldsymbol{e}_{12}$ points to the ankle 1 of standing foot.
(2) Coordinate system 2 is fixed in the hip joint of standing leg 1, and $-\boldsymbol{e}_{22}$ points to the knee joint 1 of the standing leg.
(3) Coordinate system 3 is fixed in the hip joint of the trunk, and $\boldsymbol{e}_{32}$ points to the head.
(4) Coordinate system 4 is fixed in the hip joint of swing leg 2, and $-\boldsymbol{e}_{42}$ points to knee joint 2 of the swing leg.
(5) Coordinate system 5 is fixed in knee joint 2 of swing leg 2, and $-\boldsymbol{e}_{52}$ points to ankle 2 of the swing leg.
(6) Coordinate system 6 is fixed in ankle 2 of swing foot 2, and $-\boldsymbol{e}_{61}$ points to the tiptoe of swing foot 2.

In the figure, $O_i(i = 0, \ldots, 6)$ represents the original point of each coordinate system; $\boldsymbol{e}_{ij}$ represents the unit vector expressed by the i in the coordinate system. q_i $(i = 1, \ldots, 6)$ represents each joint angle, and the counterclockwise is positive; q_i is defined as follows, respectively:

(1): q_1 represents the flexion angle of ankle joint 1.
(2): q_2 represents the flexion angle of knee joint 1.
(3): q_3 represents hip joint's stretching angle of thigh 1.
(4): q_4 represents the hip joint's flexion angle of thigh 2.
(5): q_5 represents the flexion angle of knee joint 2.
(6): q_6 represents the flexion angle of ankle joint 2.

2.2.4 Partition of the Model

According to the constraint relationship between the exoskeleton suit and the ground and at the same time considering the movement status of the legs, Racine has divided the exoskeleton suit's walking process into several main models [7], as follows:

(1) Jumping model is the state in which the feet are off the ground and do not touch the ground.
(2) Single support model is the state in which one foot touches the ground and the other foot is completely off the ground.
(3) Double support model is the state in which the two soles fully touch the ground.
(4) Double support and single redundance state is the state in which a sole fully touches the ground, and for the other one, only the toe or heel touches the ground.
(5) Double support and double redundance state is the state in which for either foot, only the toe or heel touches the ground.

When using this classification method, there are too many models, and the dynamic equation is complex. For example, the single support model as shown in Fig. 2.4, due to considering the movement of two legs at the same time, there are seven degrees of freedom (in the figure, the rotational degree of freedom of the tiptoe rotating on the ground caused by tiptoe touching the land and heel being off the land is not given).

In this book, models of each leg are classified alone, which are divided into support model and swing model. This method is more simple, for example, for the left leg, in every moment, no matter what state the right leg is in, there are only two kinds of states for the left leg, one being the support model and the other being the swing model, and there are only three degrees of freedom for each model (only considering the longitudinal plane). For the support model as shown in Fig. 2.5a,

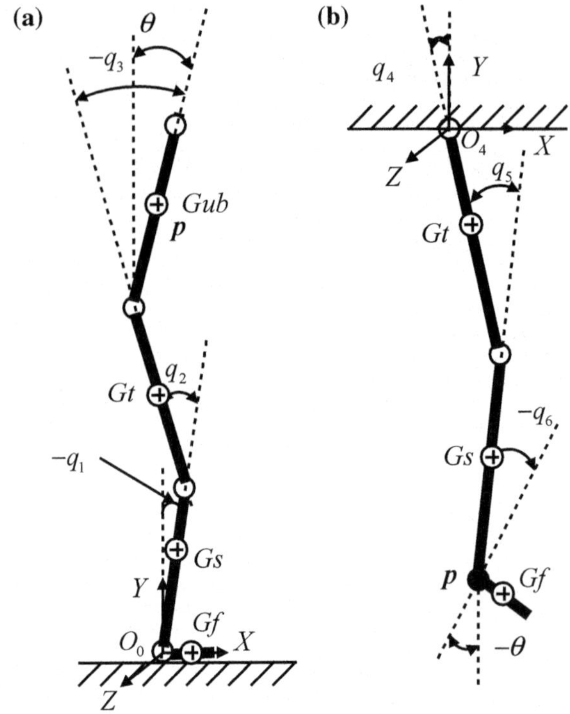

Fig. 2.5 Exoskeleton suit's model classification and the end effector's schematic diagram: **a** the support leg and **b** the swing leg

the left leg is composed of three connecting rods of shanks, thighs, and torso. For the swing model as shown in Fig. 2.5b, the left leg is composed of three connecting rods of thighs, shanks, and feet. The model classification method simplifies the exoskeleton suit's model, and for each leg's models, just switch between these two kinds.

Strictly speaking, there are mutual influence relations between two legs. For example, when the left leg is in the support model and the right leg in the swing model, all the weight of the torso is borne by the left leg in the support model; when the right leg is in the support model, the weight of the torso is borne by the two legs together. For left leg support state model, because of the change of right leg model, the relative weight of the torso changes, causing the change of model parameters, but this change can be predicted by judging the right leg model, so the influence on system model parameter can be remedied. Due to the limitation of the book's length, this book does not research the mutual influence between the two legs for the moment.

In this book, we will establish the models of exoskeleton suit's each leg separately, and we divided each leg's models into support model and swing model for analysis. The method of establishing the kinematics model and dynamics model will be introduced in the next parts.

2.3 Kinematics Model of Exoskeleton Suit

2.3.1 Position and Orientation Description of Rigid Body

Any joint connecting rod of the exoskeleton suit can be regarded as a rigid body, and the pose of the rigid body in space is made up of the rigid body's position ${}^0\boldsymbol{p} = [p_x \quad p_y \quad p_z]^{\mathrm{T}}$ and orientation in reference coordinate system $\{0\}$. The position can be represented with three-dimensional vector $\boldsymbol{p}$, namely

$$ {}^0\boldsymbol{p} = [p_x \quad p_y \quad p_z]^{\mathrm{T}} \tag{2.1} $$

And orientation can be represented with rotation matrix $\boldsymbol{R}$ from the local coordinate system $\{i\}$ fixed on the rigid body to the reference coordinate system $\{0\}$, namely

$$ {}^0_i\boldsymbol{R} = \begin{bmatrix} r_{11} & r_{12} & r_{13} \\ r_{21} & r_{22} & r_{23} \\ r_{31} & r_{32} & r_{33} \end{bmatrix} \tag{2.2} $$

The pose of any point $\boldsymbol{p}$ in the space is different for different coordinate systems, and the expression of $\boldsymbol{p}$ among different coordinate systems can be obtained by translating coordinate transformation and rotating coordinate transformation. If the orientation of coordinate system $\{i\}$ and that of coordinate system $\{j\}$ are same, but their origins are different, then the origin's vector of coordinate system $\{j\}$ in the coordinate system $\{i\}$ is ${}^i\boldsymbol{p}_{Oj}$. Assuming position vector of the point $\boldsymbol{p}$ in coordinate system $\{j\}$ is ${}^j\boldsymbol{p}$, then, its position vector ${}^i\boldsymbol{p}$ in coordinate system $\{i\}$ can be obtained by the formula

$$ {}^i\boldsymbol{p} = {}^j\boldsymbol{p} + {}^i\boldsymbol{p}_{Oj} \tag{2.3} $$

If the origin O_i of coordinate system $\{i\}$ coincides with the origin O_j of coordinate system $\{j\}$, their orientations are different, as shown in Fig. 2.6. Coordinate system $\{j\}$ can be obtained in this way: First, coordinate system $\{i\}$ revolves ψ angle around y_i-axis to obtain coordinate system $\{O_ix'y_iz'\}$, then $\{O_ix'y_iz'\}$ revolves θ angle around z'-axis to obtain coordinate system $\{O_ix_jy'z'\}$, and then $\{O_ix_jy'z'\}$ revolves γ angle around x_j-axis to obtain coordinate system $\{O_jx_jy_jz_j\}$, namely coordinate system $\{j\}$. Use the ${}^i_j\boldsymbol{R}$ representing rotation matrix of the coordinate system $\{j\}$ relative to the coordinate system $\{i\}$; use the ${}^{\;\;i}_{O_ix'y_iz'}\boldsymbol{R}$ representing rotation matrix of the coordinate system $\{O_ix'y_iz'\}$ relative to the coordinate system $\{i\}$; use the ${}^{O_ix'y_iz'}_{O_ix_jy'z'}\boldsymbol{R}$ representing rotation matrix of the coordinate system $\{O_ix_jy'z'\}$ relative to the coordinate system $\{O_ix'y_iz'\}$; use the ${}^{O_ix_jy'z'}_{\;\;j}\boldsymbol{R}$ representing rotation

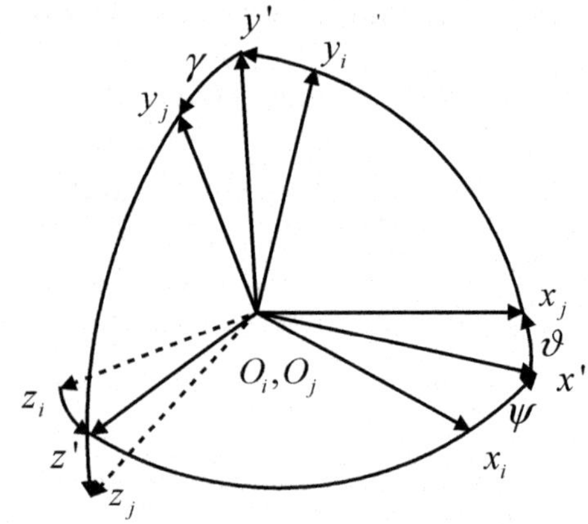

Fig. 2.6 Rotation transformation of the coordinate

matrix of the coordinate system $\{j\}$ relative to the coordinate system $\{O_i x_j y' z'\}$; then,

$$ {}_{O_i x' y_i z'}^{\ \ i}\boldsymbol{R} = \begin{bmatrix} \cos\psi & 0 & \sin\psi \\ 0 & 1 & 0 \\ -\sin\psi & 0 & \cos\psi \end{bmatrix} \tag{2.4} $$

$$ {}_{O_i x_j y' z'}^{O_i x' y_i z'}\boldsymbol{R} = \begin{bmatrix} \cos\theta & -\sin\theta & 0 \\ \sin\theta & \cos\theta & 0 \\ 0 & 0 & 1 \end{bmatrix} \tag{2.5} $$

$$ {}_{\ \ \ j}^{O_i x_j y' z'}\boldsymbol{R} = \begin{bmatrix} 1 & 0 & 0 \\ 0 & \cos\gamma & -\sin\gamma \\ 0 & \sin\gamma & \cos\gamma \end{bmatrix} \tag{2.6} $$

$$ {}_{j}^{i}\boldsymbol{R} = {}_{O_i x' y_i z'}^{\ \ i}\boldsymbol{R} \cdot {}_{O_i x_j y' z'}^{O_i x' y_i z'}\boldsymbol{R} \cdot {}_{\ \ \ j}^{O_i x_j y' z'}\boldsymbol{R} \tag{2.7} $$

Assuming position vector of the point $\boldsymbol{p}$ in coordinate system $\{j\}$ is ${}^{j}\boldsymbol{p}$, then, its position vector ${}^{i}\boldsymbol{p}$ in coordinate system $\{i\}$ can be obtained by the formula

$$ {}^{i}\boldsymbol{p} = {}_{j}^{i}\boldsymbol{R} \cdot {}^{j}\boldsymbol{p} \tag{2.8} $$

Using rotation matrix ${}_{j}^{i}\boldsymbol{R}$, three-dimensional space can be transformed from a given posture to any posture; therefore, the three variables (ψ, θ, γ) can represent any posture, often called yawing–pitching–rolling notation, also known as $X-Y-Z$ Euler angle.

For the most general case, namely, the origin of coordinate system $\{j\}$ does not coincide with that coordinate system $\{j\}$, and their orientations are different. Then, the position vector can be obtained through the composite transformation of translation transformation and rotation transformation; namely, position vector of the point $\boldsymbol{p}$ in coordinate system $\{j\}$ is ${}^{j}\boldsymbol{p}$

$$ {}^{i}\boldsymbol{p} = {}^{i}_{j}\boldsymbol{R} \cdot {}^{j}\boldsymbol{p} + {}^{i}\boldsymbol{p}_{Oj} \tag{2.9} $$

But the equation (2.9) is not homogeneous for ${}^{j}\boldsymbol{p}$ and so generally transforms the equation (2.9) into equivalent homogeneous transformation form

$$ \begin{bmatrix} {}^{i}\boldsymbol{p} \\ 1 \end{bmatrix} = \begin{bmatrix} {}^{i}_{j}\boldsymbol{R} & {}^{i}\boldsymbol{p}_{Oj} \\ 0 & 1 \end{bmatrix} \cdot \begin{bmatrix} {}^{j}\boldsymbol{p} \\ 1 \end{bmatrix} \tag{2.10} $$

Under the condition of without causing confusion, use 4×1 column vector representing point coordinates of three-dimensional coordinate system, called homogeneous coordinates, and still mark them as ${}^{i}\boldsymbol{p}$ and ${}^{i}\boldsymbol{p}$. Then, the equation (2.10) can be written in the following form:

$$ {}^{i}\boldsymbol{p} = {}^{i}_{j}\boldsymbol{T} \cdot {}^{j}\boldsymbol{p} \tag{2.11} $$

In the equation,

$$ {}^{i}_{j}\boldsymbol{T} = \begin{bmatrix} {}^{i}_{j}\boldsymbol{R} & {}^{i}\boldsymbol{p}_{Oj} \\ 0 & 1 \end{bmatrix} \tag{2.12} $$

The equation (2.12) is called the homogeneous transformation matrix. If we have known the position vector of point $\boldsymbol{p}$ in the coordinate system $\{i\}$ or $\{j\}$, then we can easily obtain the position vector in another coordinate system by homogeneous transformation equation (2.11).

2.3.2 Kinematics Model

Kinematics model describes the relationship between the space pose and joint angle. The space pose refers to the space pose of the end effector, and the space in which the space pose is located is generally referred to as the operating space, work space, or task space, and in this book, we will call the space as operating space. The space in which the joint angle is located is called the joint space. So we can say the kinematics model describes the relationship between the joint space and operating space. Generally, the robot's end effector is unique, while the exoskeleton suit's end effector is related to the system's state. If a certain leg of the exoskeleton suit is in the support model, then the foot-supporting point of exoskeleton suit is regarded as the base coordinate, and the trunk's center of gravity of the exoskeleton suit is regarded as the end effector of the supporting leg connecting rod, as shown in

Fig. 2.5a. If a certain leg of the exoskeleton suit is in the swing model, then the hip joint of the leg is regarded as the base coordinate, and the ankle of exoskeleton suit is regarded as the end effector of the swing leg, as shown in Fig. 2.5a. In the next part, we will analyze the kinematics of exoskeleton suit based on the position and speed relationship of operating space and the joint space, respectively.

The position relationship

We take the left supporting leg as the example to make the kinematics analysis. Refer to coordinate system definition in Fig. 2.4: The origin of reference coordinate system $\{0\}$ is defined as the point O_0, and then, the position vector of the left knee joint point O_1 in the reference coordinate system $\{0\}$ is

$$^{0}\boldsymbol{p}_{O1} = [-L_s \sin q_1 \quad L_s \cos q_1 \quad 0]^{\mathrm{T}} \tag{2.13}$$

The rotation transformation matrix from left knee joint coordinate system $\{1\}$ to the reference coordinate system $\{0\}$ is

$$^{0}_{1}\boldsymbol{R} = \begin{bmatrix} \cos q_1 & -\sin q_1 & 0 \\ \sin q_1 & \cos q_1 & 0 \\ 0 & 0 & 1 \end{bmatrix} \tag{2.14}$$

Then, the homogeneous transformation matrix from coordinate system $\{1\}$ to the reference coordinate system $\{0\}$ is as follows:

$$^{0}_{1}\boldsymbol{T} = \begin{bmatrix} ^{0}_{1}\boldsymbol{R} & ^{0}\boldsymbol{p}_{O1} \\ 0 & 1 \end{bmatrix} = \begin{bmatrix} \cos q_1 & -\sin q_1 & 0 & -L_s \sin q_1 \\ \sin q_1 & \cos q_1 & 0 & L_s \cos q_1 \\ 0 & 0 & 1 & 0 \\ 0 & 0 & 0 & 1 \end{bmatrix} \tag{2.15}$$

In the same way, the homogeneous transformation matrix from left hip joint coordinate system $\{2\}$ to the left knee joint coordinate system $\{1\}$ is as follows:

$$^{1}_{2}\boldsymbol{T} = \begin{bmatrix} ^{1}_{2}\boldsymbol{R} & ^{1}\boldsymbol{p}_{O2} \\ 0 & 1 \end{bmatrix} = \begin{bmatrix} \cos q_2 & -\sin q_2 & 0 & -L_t \sin q_2 \\ \sin q_2 & \cos q_2 & 0 & L_t \cos q_2 \\ 0 & 0 & 1 & 0 \\ 0 & 0 & 0 & 1 \end{bmatrix} \tag{2.16}$$

The origin O_3 of the trunk coordinate system $\{3\}$ coincides with the origin O_2 of the left hip joint coordinate system $\{2\}$, so

$$^{2}_{3}\boldsymbol{T} = \begin{bmatrix} ^{2}_{3}\boldsymbol{R} & ^{2}\boldsymbol{p}_{O3} \\ 0 & 1 \end{bmatrix} = \begin{bmatrix} \cos q_3 & -\sin q_3 & 0 & 0 \\ \sin q_3 & \cos q_3 & 0 & 0 \\ 0 & 0 & 1 & 0 \\ 0 & 0 & 0 & 1 \end{bmatrix} \tag{2.17}$$

The vector of trunk's center of gravity in trunk coordinate system $\{3\}$ can be expressed as

$$^{2}\boldsymbol{p}_{\text{Gub}} = \begin{bmatrix} -L_{\text{Gub}}\sin q_3 & L_{\text{Gub}}\cos q_3 & 0 \end{bmatrix}^{\text{T}} \tag{2.18}$$

The homogeneous transformation matrix from trunk's center of gravity to the left hip joint coordinate system is as follows:

$$^{2}_{3}\boldsymbol{T}_{\text{Gub}} = \begin{bmatrix} ^{2}_{3}\boldsymbol{R} & ^{2}\boldsymbol{p}_{\text{Gub}} \\ 0 & 1 \end{bmatrix} = \begin{bmatrix} \cos q_3 & -\sin q_3 & 0 & -L_{\text{Gub}}\sin q_3 \\ \sin q_3 & \cos q_3 & 0 & L_{\text{Gub}}\cos q_3 \\ 0 & 0 & 1 & 0 \\ 0 & 0 & 0 & 1 \end{bmatrix} \tag{2.19}$$

So, according to the chain rule of the homogeneous transformation matrix, the homogeneous transformation matrix from trunk's center of gravity to reference coordinate system is as follows:

$$^{0}_{3}\boldsymbol{T}_{\text{Gub}} = {}^{0}_{1}\boldsymbol{T} \cdot {}^{1}_{2}\boldsymbol{T} \cdot {}^{2}_{3}\boldsymbol{T}_{\text{Gub}} \tag{2.20}$$

Making use of symbolic operation toolbox of MATLAB, we can calculate to obtain the $^{0}_{3}\boldsymbol{T}_{\text{Gub}}$ quickly:

$$^{0}_{3}\boldsymbol{T}_{\text{Gub}} = \begin{bmatrix} ^{0}_{3}\boldsymbol{R} & ^{0}\boldsymbol{p}_{\text{Gub}} \\ 0 & 1 \end{bmatrix} \tag{2.21}$$

In the equation,

$$^{0}_{3}\boldsymbol{R} = \begin{bmatrix} \cos(q_1+q_2+q_3) & -\sin(q_1+q_2+q_3) & 0 \\ \sin(q_1+q_2+q_3) & \cos(q_1+q_2+q_3) & 0 \\ 0 & 0 & 1 \end{bmatrix} \tag{2.22}$$

$$^{0}\boldsymbol{p}_{\text{Gub}} = \begin{bmatrix} -L_s\sin q_1 - L_t\sin(q_2+q_1) - L_{\text{Gub}}\sin(q_3+q_2+q_1) \\ L_s\cos q_1 + L_t\cos(q_2+q_1) + L_{\text{Gub}}\cos(q_3+q_2+q_1) \\ 0 \end{bmatrix} \tag{2.23}$$

Making use of the same method, we can obtain the homogeneous transformation matrix from the swing leg's ankle joint to hip joint coordinate system.

2.3.2.1 Speed Relationship

If we define the $\dot{\boldsymbol{q}}$ to represent the exoskeleton suit's joint velocity, define the $\dot{\boldsymbol{p}}$ to represent 3×1 translational velocity vector of trunk's center of gravity of exoskeleton suit relative to reference coordinate system, and define $\boldsymbol{\omega}$ to represent the 3×1 rotation angular velocity vector of trunk's center of gravity of exoskeleton suit relative to reference coordinate system, then the velocity of trunk's center of gravity in reference coordinate system can be expressed as

$$\boldsymbol{v} = \begin{bmatrix} \dot{\boldsymbol{p}} \\ \boldsymbol{\omega} \end{bmatrix} \tag{2.24}$$

And there is such relationship between $\boldsymbol{v}$ and $\dot{\boldsymbol{q}}$:

$$\boldsymbol{v} = \boldsymbol{J}(\boldsymbol{q}) \cdot \dot{\boldsymbol{q}} \tag{2.25}$$

In the equation, $\boldsymbol{J}$ is a 6×3 matrix, representing the geometric Jacobian matrix of the trunk's center of gravity of exoskeleton suit, and $\boldsymbol{J}$ can be divided into two parts, namely

$$\boldsymbol{J} = \begin{bmatrix} \boldsymbol{J}_{\mathrm{p}} \\ \boldsymbol{J}_{\omega} \end{bmatrix} \tag{2.26}$$

The two parts correspond to the translational velocity and angular velocity In equation (2.24) respectively.

We take the derivative of the equation (2.24), and then, we can obtain

$$\boldsymbol{J}_{\mathrm{p}} = [\boldsymbol{J}_{\mathrm{p1}} \quad \boldsymbol{J}_{\mathrm{p2}} \quad \boldsymbol{J}_{\mathrm{p3}}]^{\mathrm{T}} \tag{2.27}$$

In the equation,

$$\boldsymbol{J}_{\mathrm{p1}} = \begin{bmatrix} -L_{\mathrm{s}} \cos q_1 - L_{\mathrm{t}} \cos(q_2 + q_1) - L_{\mathrm{Gub}} \cos(q_3 + q_2 + q_1) \\ -L_{\mathrm{t}} \cos(q_2 + q_1) - L_{\mathrm{Gub}} \cos(q_3 + q_2 + q_1) \\ -L_{\mathrm{Gub}} \cos(q_3 + q_2 + q_1) \end{bmatrix}^{\mathrm{T}} \tag{2.28}$$

$$\boldsymbol{J}_{\mathrm{p2}} = \begin{bmatrix} -L_{\mathrm{s}} \sin q_1 - L_{\mathrm{t}} \sin(q_2 + q_1) - L_{\mathrm{Gub}} \sin(q_3 + q_2 + q_1) \\ -L_{\mathrm{t}} \sin(q_2 + q_1) - L_{\mathrm{Gub}} \sin(q_3 + q_2 + q_1) \\ -L_{\mathrm{Gub}} \sin(q_3 + q_2 + q_1) \end{bmatrix}^{\mathrm{T}} \tag{2.29}$$

$$\boldsymbol{J}_{\mathrm{p3}} = [0 \quad 0 \quad 0] \tag{2.30}$$

There are such relationships among the angular velocity $\boldsymbol{\omega}$, the rotation matrix, and its differential:

$${}_{3}^{0}\dot{\boldsymbol{R}} = \boldsymbol{S}(\boldsymbol{\omega}) \cdot {}_{3}^{0}\boldsymbol{R} \tag{2.31}$$

If the $\boldsymbol{\omega}$ can be expressed as follows:

$$\boldsymbol{\omega} = \begin{bmatrix} \omega_x & \omega_y & \omega_z \end{bmatrix}^{T} \tag{2.32}$$

Then, $\boldsymbol{S}(\boldsymbol{\omega})$ can be defined as follows:

$$\boldsymbol{S} = \begin{bmatrix} 0 & -\omega_z & \omega_y \\ \omega_z & 0 & -\omega_x \\ -\omega_y & \omega_x & 0 \end{bmatrix} \tag{2.33}$$

And $\boldsymbol{S}(\boldsymbol{\omega})$ is skew symmetric matrix, namely

$$\boldsymbol{S}^{T}(\boldsymbol{\omega}) = -\boldsymbol{S}(\boldsymbol{\omega}) \tag{2.34}$$

By the equation (2.31), we can obtain

$$\boldsymbol{S}(\boldsymbol{\omega}) = {}^{0}_{3}\dot{\boldsymbol{R}} \cdot {}^{0}_{3}\boldsymbol{R}^{T} \tag{2.35}$$

Equation (2.22) differentiates the time, and then, the following can be obtained:

$${}^{0}_{3}\dot{\boldsymbol{R}} = (\dot{q}_1 + \dot{q}_2 + \dot{q}_3) \cdot \begin{bmatrix} -\sin(q_1 + q_2 + q_3) & -\cos(q_1 + q_2 + q_3) & 0 \\ \cos(q_1 + q_2 + q_3) & -\sin(q_1 + q_2 + q_3) & 0 \\ 0 & 0 & 0 \end{bmatrix} \tag{2.36}$$

So, by the equations (2.22), (2.35) and (2.36), we can obtain

$$\boldsymbol{S}(\boldsymbol{\omega}) = \begin{bmatrix} 0 & -(\dot{q}_1 + \dot{q}_2 + \dot{q}_3) & 0 \\ \dot{q}_1 + \dot{q}_2 + \dot{q}_3 & 0 & 0 \\ 0 & 0 & 0 \end{bmatrix} \tag{2.37}$$

Contrasting Eq. (2.33) with Eq. (2.37), we can obtain

$$\begin{aligned} \omega_x &= 0 \\ \omega_y &= 0 \\ \omega_z &= \dot{q}_1 + \dot{q}_2 + \dot{q}_3 \end{aligned} \tag{2.38}$$

Namely,

$$\boldsymbol{\omega} = \begin{bmatrix} 0 & 0 & \dot{q}_1 + \dot{q}_2 + \dot{q}_3 \end{bmatrix}^{T} \tag{2.39}$$

So,

$$\boldsymbol{J}_{\omega} = \begin{bmatrix} \boldsymbol{J}_{\omega 1} \\ \boldsymbol{J}_{\omega 2} \\ \boldsymbol{J}_{\omega 3} \end{bmatrix} = \begin{bmatrix} 0 & 0 & 0 \\ 0 & 0 & 0 \\ 1 & 1 & 1 \end{bmatrix} \tag{2.40}$$

After observing Eqs. (2.23), (2.39), we find that because of only considering the longitudinal plane, the position of trunk's center of gravity in the z-axis direction in the reference coordinate system is always zero. And the rotation angular velocity of the trunk revolving around x-axis and y-axis is also zero; therefore, we can use the way of dimension reduction to show the generalized position of exoskeleton suit's trunk in the operating space, namely defining the generalized coordinates of the exoskeleton suit's trunk:

$$\boldsymbol{x} = \begin{bmatrix} \boldsymbol{p} \\ \theta \end{bmatrix} \tag{2.41}$$

In the equation, $\boldsymbol{p} = [p_{\mathrm{x}} \quad p_{\mathrm{y}}]^{\mathrm{T}} \in \mathbb{R}^2$ is the plane position vector of the trunk's center of gravity of exoskeleton suit in the operating space, and by Eq. (2.23), we can obtain

$$p_{\mathrm{x}} = -L_{\mathrm{s}} \sin q_1 - L_{\mathrm{t}} \sin(q_2 + q_1) - L_{\mathrm{Gub}} \sin(q_3 + q_2 + q_1) \tag{2.42}$$

$$p_{\mathrm{y}} = L_{\mathrm{s}} \cos q_1 + L_{\mathrm{t}} \cos(q_2 + q_1) + L_{\mathrm{Gub}} \cos(q_3 + q_2 + q_1) \tag{2.43}$$

And the $\theta = q_1 + q_2 + q_3$ represents the orientation of the exoskeleton suit's trunk, and then, $\dot{\theta} = \omega_{\mathrm{z}}$. According to Eq. (2.41), the generalized velocity of exoskeleton suit's trunk is as follows:

$$\dot{\boldsymbol{x}} = \begin{bmatrix} \dot{\boldsymbol{p}} \\ \dot{\theta} \end{bmatrix} \tag{2.44}$$

According to Eqs. (2.25), (2.28), and (2.29), we obtain the equation:

$$\dot{\boldsymbol{x}} = \boldsymbol{J}(\boldsymbol{q})\dot{\boldsymbol{q}} \tag{2.45}$$

In the equation, $\boldsymbol{J}(\boldsymbol{q})$ is the Jacobian matrix of 3×3, and

$$\boldsymbol{J}(\boldsymbol{q}) = \begin{bmatrix} \boldsymbol{J}_{\mathrm{p}1} \\ \boldsymbol{J}_{\mathrm{p}2} \\ \boldsymbol{J}_{\omega 3} \end{bmatrix} \tag{2.46}$$

Similarly, for the swing leg, as shown in Fig. 2.5b, also we can obtain

$$\boldsymbol{x} = \begin{bmatrix} \boldsymbol{p} \\ \theta \end{bmatrix} = \begin{bmatrix} p_x \\ p_y \\ \theta \end{bmatrix} = \begin{bmatrix} L_t \sin q_1 + L_s \sin(q_1 + q_2) \\ L_t \cos q_1 + L_s \cos(q_1 + q_2) \\ q_1 + q_2 + q_3 \end{bmatrix} \tag{2.47}$$

Taking the (2.47) in the form of partial differential equation, we can obtain the Jacobian matrix

$$\boldsymbol{J}(\boldsymbol{q}) = \begin{bmatrix} L_t \cos q_1 + L_s \cos(q_1 + q_2) & L_s \cos(q_1 + q_2) & 0 \\ -L_t \sin q_1 - L_s \sin(q_1 + q_2) & -L_s \sin(q_1 + q_2) & 0 \\ 1 & 1 & 1 \end{bmatrix} \tag{2.48}$$

In the actual movement, in some certain condition, the exoskeleton suit's pose can lead to Jacobian matrix's module $|\boldsymbol{J}| = 0$, and namely, singular value exists. In the singular value, the Jacobian matrix fails, therefore, in the control of exoskeleton suit, and the occurrence of singular pose should be tried to avoid.

Equations (2.41) and (2.45) are the expression of exoskeleton suit's kinematics model. The kinematics model describes the relationship between the joint space and the operation space. In this book, Chap. 3 discusses the questions in joint space, and from Chaps. 4–7, questions are discussed in joint space and operation space at the same time.

2.4 Dynamics of Exoskeleton Suit

Dynamics modeling is conducted on the basis of kinematics modeling, which mainly researches the force/torque's acting relationship among all the system's parts in the process of walking, to determine the driving torque of each joint on the basis of the drive and control. The exoskeleton suit corresponds to different models under the different movement models, so in the dynamic modeling, we need to make modeling analysis, respectively, according to different models.

2.4.1 The Method of Dynamics Modeling

For the multi-rigid-body system, there are a lot of dynamic modeling methods. Euler–Lagrange method is more mature classical method of multi-rigid-body dynamic modeling [8]. Euler–Lagrange method is the method that differentiates the system variables and time based on the energy term. For simple cases, using this method is more complicated than using Newton's mechanics method; however, with the increase of system's complexity, using the Euler–Lagrange method becomes relatively easier. The Euler–Lagrange method is based on the following two basic equations: one for rectilinear motion and the other for rotary motion. First of all, define the Lagrange functions as

$$L = KE - V \tag{2.49}$$

In the equation, *KE* is system dynamics; *V* is potential energy of the system. So

$$\boldsymbol{T} = \frac{\partial}{\partial t}\frac{\partial L}{\partial \dot{\boldsymbol{q}}} - \frac{\partial L}{\partial \boldsymbol{q}} \tag{2.50}$$

In the equation, $\boldsymbol{T}$ is all the resultant external moment vector in the movement; $\boldsymbol{q}$ is the system's joint angle motion vector. The other method widely used in robot dynamics is KANE method [9, 10], and it can also be applied to establish dynamic equation of a multi-body system. As same as the Lagrange method, KANE method is also a standard method to write dynamic equation. The concrete form is as follows:

$$\boldsymbol{K}_{\mathrm{r}} + \boldsymbol{K}_{\mathrm{r}}^{*} = 0 \tag{2.51}$$

In the equation, $\boldsymbol{K}_{\mathrm{r}}$ and $\boldsymbol{K}_{\mathrm{r}}^{*}$ are generalized active force and generalized inertia force, respectively.

Comparing the above two methods, their main differences are as follows: For the mechanical system of n degrees of freedom, Lagrange equation gives n second-order differential equations, and KANE method gives the $2n$ first-order differential equation. The Lagrange equation only need write the system's kinetic energy and potential energy shown in generalized coordinates, and the concept is clear, while the KANE method must construct appropriate generalized velocity and partial velocity. In fact, when adopting the generalized coordinates adopted by Lagrange, the models derived by the two methods are equivalent [11].

Another method is the R/W method proposed by Roberson of University of San Diego in the USA and Wittenburg of University of Karlsruhekars in Germany. They have introduced the graph theory in mathematics into dynamics to study the safety problem of car crash [12]. Because this method is more complex, it is used less.

In this book, we use Euler–Lagrange method to make a dynamic modeling analysis of the exoskeleton suit.

2.4.2 Modeling Procedures

The modeling method of support state model and swing state model is the same. That is, firstly calculate the kinetic energy and potential energy to obtain the Lagrange function shown in Eq. (2.49), and then, calculate to obtain the corresponding dynamic model by differential operation shown in Eq. (2.50). The following takes the example of the steps of modeling the support state model to explain [7].

2.4.2.1 The Unit Vector

If we define $\boldsymbol{e}_{ij/k}$ to represent the unit vector $\boldsymbol{e}_{ij}$ expressed in the coordinate system k, as shown in Fig. 2.4, then in the inertial coordinate system, the unit vector can be expressed as

$$\begin{aligned} \boldsymbol{e}_{01/0} &= \begin{bmatrix} 1 & 0 & 0 \end{bmatrix}^{\mathrm{T}} \\ \boldsymbol{e}_{02/0} &= \begin{bmatrix} 0 & 1 & 0 \end{bmatrix}^{\mathrm{T}} \\ \boldsymbol{e}_{03/0} &= \begin{bmatrix} 0 & 0 & 1 \end{bmatrix}^{\mathrm{T}} \end{aligned} \tag{2.52}$$

By Eq. (2.8), we can know that the vector $\boldsymbol{e}_{/i}$ expressed in the coordinate system i can be expressed as the vector in the coordinate system j by transforming $\boldsymbol{e}_{/j} = {}_{i}^{j}\boldsymbol{R} \cdot \boldsymbol{e}_{/i}$. Therefore, $\forall i \in [1,2,3]$ and $\forall j \in [1,2,3]$, and then, the unit vector of each coordinate system in the inertial coordinate system can be expressed as

$$\boldsymbol{e}_{ij/0} = {}_{i}^{0}\boldsymbol{R} \cdot \boldsymbol{e}_{ij/i} \tag{2.53}$$

2.4.2.2 The Position Vector

Each key position vector is transformed into the position vector in the reference coordinate system, and let the $\boldsymbol{r}_{O1O2/0}$ express the expression of the vector from point $O1$ to point $O2$ in the coordinate system $\{0\}$ (the reference coordinate system), and the rest can be done in the same manner; so there are the following:

The vector from the ankle to the shank's center of gravity is

$$\boldsymbol{r}_{O0\mathrm{Gs}/0} = L_{\mathrm{Gs}}\boldsymbol{e}_{12/0} \tag{2.54}$$

The vector from the ankle to the knee joint is

$$\boldsymbol{r}_{O0O1/0} = L_{\mathrm{s}}\boldsymbol{e}_{12/0} \tag{2.55}$$

The vector from the knee joint to thigh's center of gravity is

$$\boldsymbol{r}_{O1\mathrm{Gt}/0} = L_{\mathrm{Gt}}\boldsymbol{e}_{22/0} \tag{2.56}$$

The vector from the knee joint to the hip joint is

$$\boldsymbol{r}_{O1O2/0} = L_{\mathrm{t}}\boldsymbol{e}_{22/0} \tag{2.57}$$

The vector from the hip joint to the vector of trunk's center of gravity is

$$\boldsymbol{r}_{O3\mathrm{Gub}/0} = L_{\mathrm{Gub}}\boldsymbol{e}_{32/0} \tag{2.58}$$

2.4.2.3 Link Angular Velocity

ω_{ij} expresses the rotating angular velocity vector of link i relative to link j, and link 0 expresses the reference coordinate system. Then, for given $i \in [1,2,3]$, the angular velocity of two adjacent links can be expressed as

$$\boldsymbol{\omega}_{i(i-1)} = \dot{q}_i \boldsymbol{e}_{03/0} \tag{2.59}$$

For given $i \in [2,3]$, the angular velocity of each link relative to the reference coordinate system can be expressed as

$$\boldsymbol{\omega}_{i0} = \boldsymbol{\omega}_{i(i-1)} + \boldsymbol{\omega}_{(i-1)0} \tag{2.60}$$

2.4.2.4 The Velocity About the Point

Let $\boldsymbol{v}_{O/0}$ express the linear velocity of the point O in the reference coordinate system, and the rest can be done in the same manner; so there are the following:

The linear velocity of shank's center of gravity is

$$\boldsymbol{v}_{\mathrm{Gs}/0} = \boldsymbol{\omega}_{20} \times \boldsymbol{r}_{O0\mathrm{Gs}/0} \tag{2.61}$$

The linear velocity of the knee joint is

$$\boldsymbol{v}_{O1/0} = \boldsymbol{\omega}_{10} \times \boldsymbol{r}_{O0O1/0} \tag{2.62}$$

The linear velocity of thigh's center of gravity is

$$\boldsymbol{v}_{\mathrm{Gt}/0} = \boldsymbol{v}_{O1/0} + \boldsymbol{\omega}_{20} \times \boldsymbol{r}_{O1\mathrm{Gt}/0} \tag{2.63}$$

The linear velocity of hip joint is

$$\boldsymbol{v}_{O2/0} = \boldsymbol{v}_{O1/0} + \boldsymbol{\omega}_{20} \times \boldsymbol{r}_{O1O2/0} \tag{2.64}$$

The linear velocity of trunk's gravity is

$$\boldsymbol{v}_{\mathrm{Gub}/0} = \boldsymbol{v}_{O2/0} + \boldsymbol{\omega}_{30} \times \boldsymbol{r}_{O3\mathrm{Gub}/0} \tag{2.65}$$

2.4.2.5 The System's Kinetic Energy

The shank's kinetic energy is

$$\mathrm{KE}_{\mathrm{s}} = \frac{1}{2} m_{\mathrm{s}} \boldsymbol{v}_{\mathrm{Gs}/0} \boldsymbol{v}_{\mathrm{Gs}/0} + \frac{1}{2} I_{\mathrm{s}} \boldsymbol{\omega}_{10} \boldsymbol{\omega}_{10} \tag{2.66}$$

The thigh's kinetic energy is

$$\mathrm{KE_t} = \frac{1}{2} m_t \boldsymbol{v}_{\mathrm{Gt}/0} \boldsymbol{v}_{\mathrm{Gt}/0} + \frac{1}{2} I_t \omega_{20} \omega_{20} \tag{2.67}$$

The trunk's kinetic energy is

$$\mathrm{KE_{ub}} = \frac{1}{2} m_{\mathrm{ub}} \boldsymbol{v}_{\mathrm{Gub}/0} \boldsymbol{v}_{\mathrm{Gub}/0} + \frac{1}{2} I_{\mathrm{ub}} \omega_{40} \omega_{40} \tag{2.68}$$

If the other leg is also in the support state, then the trunk's weight and rotary inertia are half of that of the current. The total kinetic energy of the system is

$$\mathrm{KE} = \mathrm{KE_s} + \mathrm{KE_t} + \mathrm{KE_{ub}} \tag{2.69}$$

2.4.2.6 The System's Potential Energy

The shank's potential energy is

$$V_s = m_s g \boldsymbol{r}_{O0\mathrm{Gs}/0} \boldsymbol{e}_{02/0} \tag{2.70}$$

The thigh's potential energy is

$$V_t = m_t g (\boldsymbol{r}_{O0O1/0} + \boldsymbol{r}_{O1\mathrm{Gt}/0}) \cdot \boldsymbol{e}_{02/0} \tag{2.71}$$

The trunk's potential energy is

$$V_{\mathrm{ub}} = m_{\mathrm{ub}} g (\boldsymbol{r}_{O0O1/0} + \boldsymbol{r}_{O1O2/0} + \boldsymbol{r}_{O3\mathrm{Gub}/0}) \cdot \boldsymbol{e}_{02/0} \tag{2.72}$$

The total potential energy is

$$V = V_s + V_t + V_{\mathrm{ub}} \tag{2.73}$$

2.4.3 Dynamics Model

By Eqs. (2.69) and (2.73), we can obtain the Lagrange function L expressed by Eq. (2.49). So, according to Eq. (2.50), the system's dynamic model can be determined as

$$\boldsymbol{H}(\boldsymbol{q})\ddot{q} + \boldsymbol{C}(\boldsymbol{q}, \dot{\boldsymbol{q}})\dot{q} + \boldsymbol{G}(\boldsymbol{q}) = \boldsymbol{T} \tag{2.74}$$

In the equation, $\boldsymbol{q} = [q_1 \quad q_2 \quad q_3]^{\mathrm{T}}$; $\boldsymbol{H}(\boldsymbol{q})$ is the inertia matrix; $\boldsymbol{C}(\boldsymbol{q}, \dot{\boldsymbol{q}})$ is the Coriolis term; $\boldsymbol{G}(\boldsymbol{q})$ is the gravity term; $\boldsymbol{T} = [T_1 \quad T_2 \quad T_3]$ expresses the resultant external moment acting on exoskeleton suit, T_1 expresses ankle joint torque, T_2 expresses the knee joint torque, and T_3 expresses the hip joint torque. The concrete forms of $\boldsymbol{H}(\boldsymbol{q})$, $\boldsymbol{C}(\boldsymbol{q},\dot{\boldsymbol{q}})$, and $\boldsymbol{G}(\boldsymbol{q})$ are as follows:

$$\boldsymbol{H}(\boldsymbol{q}) = \begin{bmatrix} H_{11}(\boldsymbol{q}) & H_{12}(\boldsymbol{q}) & H_{13}(\boldsymbol{q}) \\ H_{21}(\boldsymbol{q}) & H_{22}(\boldsymbol{q}) & H_{23}(\boldsymbol{q}) \\ H_{31}(\boldsymbol{q}) & H_{32}(\boldsymbol{q}) & H_{33}(\boldsymbol{q}) \end{bmatrix} \tag{2.75}$$

$$\boldsymbol{C}(\boldsymbol{q},\dot{\boldsymbol{q}}) = \begin{bmatrix} C_{11}(\boldsymbol{q},\dot{\boldsymbol{q}}) & C_{12}(\boldsymbol{q},\dot{\boldsymbol{q}}) & C_{13}(\boldsymbol{q},\dot{\boldsymbol{q}}) \\ C_{21}(\boldsymbol{q},\dot{\boldsymbol{q}}) & C_{22}(\boldsymbol{q},\dot{\boldsymbol{q}}) & C_{23}(\boldsymbol{q},\dot{\boldsymbol{q}}) \\ C_{31}(\boldsymbol{q},\dot{\boldsymbol{q}}) & C_{32}(\boldsymbol{q},\dot{\boldsymbol{q}}) & C_{33}(\boldsymbol{q},\dot{\boldsymbol{q}}) \end{bmatrix} \tag{2.76}$$

$$G(\boldsymbol{q}) = \begin{bmatrix} G_1(\boldsymbol{q}) \\ G_2(\boldsymbol{q}) \\ G_3(\boldsymbol{q}) \end{bmatrix} \tag{2.77}$$

$$\begin{aligned} H_{11}(\boldsymbol{q}) = {} & I_t + I_{ub} + m_s L_{Gs}^2 + m_t L_s^2 + m_t L_{Gt}^2 + m_{ub} L_s^2 + m_{ub} L_t^2 \\ & + m_{ub} L_{Gub}^2 + 2 m_t L_{Gt} L_s \cos(q_2) + 2 m_{ub} L_t L_s \cos(q_2) \\ & + 2 m_{ub} L_{Gub} L_t \cos(q_3) + 2 m_{ub} L_{Gub} L_s \cos(q_2 + q_3) \end{aligned} \tag{2.78}$$

$$\begin{aligned} H_{12}(\boldsymbol{q}) = {} & I_t + I_{ub} + m_t L_{Gt}^2 + m_{ub} L_t^2 + m_{ub} L_{Gub}^2 + 2 m_{ub} L_{Gub} L_t \cos(q_3) \\ & + m_{ub} L_t \cos(q_2) + m_t L_{Gt} L_s \cos(q_2) + m_{ub} L_{Gub} L_s \cos(q_2 + q_3) \end{aligned} \tag{2.79}$$

$$H_{13}(\boldsymbol{q}) = I_u + m_{ub} L_{Gub}^2 + m_{ub} L_{Gub} L_t \cos(q_3) + m_{ub} L_{Gub} L_s \cos(q_2 + q_3) \tag{2.80}$$

$$\begin{aligned} H_{21}(\boldsymbol{q}) = {} & I_t + I_{ub} + m_t L_{Gt}^2 + m_{ub} L_{Gub}^2 + m_{ub} L_t^2 + m_t L_{Gt} L_s \cos(q_2) \\ & + m_{ub} L_s L_t \cos(q_2) + 2 m_{ub} L_{Gub} L_t \cos(q_3) \\ & + m_{ub} L_s L_{Gub} \cos(q_2 + q_3) \end{aligned} \tag{2.81}$$

$$\begin{aligned} H_{22}(\boldsymbol{q}) = {} & I_t + I_{ub} + m_{ub} L_t^2 + m_t L_{Gt}^2 + m_{ub} L_{Gub}^2 \\ & + 2 m_{ub} L_{Gub} L_t \cos(q_3) \end{aligned} \tag{2.82}$$

$$H_{23}(\boldsymbol{q}) = I_{ub} + m_{ub} L_{Gub}^2 + m_{ub} L_{Gub} L_t \cos(q_3) \tag{2.83}$$

$$\begin{aligned} H_{31}(\boldsymbol{q}) = {} & I_{ub} + m_{ub} L_{Gub}^2 + m_{ub} L_{Gub} L_t \cos(q_3) \\ & + m_{ub} L_{Gub} L_s \cos(q_2 + q_3) \end{aligned} \tag{2.84}$$

$$H_{32}(\boldsymbol{q}) = I_{ub} + m_{ub} L_{Gub}^2 + m_{ub} L_{Gub} L_t \cos(q_3) \tag{2.85}$$

$$H_{33}(\boldsymbol{q}) = I_{ub} + m_{ub} L_{Gub}^2 \tag{2.86}$$

$$\begin{aligned} C_{11}(\boldsymbol{q}, \dot{\boldsymbol{q}}) = & -2m_{\text{ub}}L_{\text{t}}L_{\text{Gub}}\dot{q}_3 \sin(q_3) - 2m_{\text{ub}}L_{\text{s}}L_{\text{t}}\dot{q}_2 \sin(q_2) \\ & - 2m_{\text{ub}}L_{\text{Gub}}\dot{q}_2L_{\text{s}} \sin(q_2 + q_3) - 2m_{\text{t}}L_{\text{Gt}}\dot{q}_2L_{\text{s}} \sin(q_2) \end{aligned} \tag{2.87}$$

$$\begin{aligned} C_{12}(\boldsymbol{q}, \dot{\boldsymbol{q}}) = & -m_{\text{ub}}L_{\text{Gub}}\dot{q}_2L_{\text{s}} \sin(q_2 + q_3) - m_{\text{ub}}L_{\text{t}}\dot{q}_2L_{\text{s}} \sin(q_2) \\ & - m_{\text{t}}L_{\text{Gt}}\dot{q}_2L_{\text{s}} \sin(q_2) - 2m_{\text{ub}}L_{\text{Gub}}\dot{q}_3L_{\text{s}} \sin(q_2 + q_3) \end{aligned} \tag{2.88}$$

$$\begin{aligned} C_{13}(\boldsymbol{q}, \dot{\boldsymbol{q}}) = & -m_{\text{ub}}L_{\text{Gub}}\dot{q}_3L_{\text{s}} \sin(q_2 + q_3) - m_{\text{ub}}L_{\text{Gub}}\dot{q}_3L_{\text{t}} \sin(q_3) \\ & - 2m_{\text{ub}}L_{\text{t}}\dot{q}_2L_{\text{Gub}} \sin(q_3) - 2m_{\text{ub}}L_{\text{Gub}}L_{\text{s}}\dot{q}_1 \sin(q_2 + q_3) \end{aligned} \tag{2.89}$$

$$\begin{aligned} C_{21}(\boldsymbol{q}, \dot{\boldsymbol{q}}) = & \, m_{\text{ub}}L_{\text{s}}\dot{q}_1L_{\text{t}} \sin(q_2) + m_{\text{ub}}L_{\text{s}}\dot{q}_1L_{\text{Gub}} \sin(q_2 + q_3) \\ & + m_{\text{t}}L_{\text{Gt}}\dot{q}_1L_{\text{s}} \sin(q_2) \end{aligned} \tag{2.90}$$

$$C_{22}(\boldsymbol{q}, \dot{\boldsymbol{q}}) = -2m_{\text{ub}}L_{\text{Gub}}\dot{q}_3L_{\text{t}} \sin(q_3) \tag{2.91}$$

$$C_{23}(\boldsymbol{q}, \dot{\boldsymbol{q}}) = -m_{\text{ub}}L_{\text{Gub}}\dot{q}_3L_{\text{t}} \sin(q_3) - 2m_{\text{ub}}L_{\text{Gub}}\dot{q}_1L_{\text{t}} \sin(q_3) \tag{2.92}$$

$$C_{31}(\boldsymbol{q}, \dot{\boldsymbol{q}}) = m_{\text{ub}}L_{\text{Gub}}\dot{q}_1L_{\text{s}} \sin(q_2 + q_3) + m_{\text{ub}}L_{\text{Gub}}L_{\text{t}}\dot{q}_1 \sin(q_3) \tag{2.93}$$

$$C_{32}(\boldsymbol{q}, \dot{\boldsymbol{q}}) = m_{\text{ub}}L_{\text{Gub}}L_{\text{t}}\dot{q}_2 \sin(q_3) + 2m_{\text{ub}}L_{\text{Gub}}L_{\text{t}}\dot{q}_1 \sin(q_3) \tag{2.94}$$

$$C_{33}(\boldsymbol{q}, \dot{\boldsymbol{q}}) = 0 \tag{2.95}$$

$$\begin{aligned} G_1(\boldsymbol{q}) = & -m_{\text{ub}}gL_{\text{s}} \sin(q_1) - m_{\text{ub}}gL_{\text{t}} \sin(q_1 + q_2) \\ & - m_{\text{ub}}gL_{\text{Gub}} \sin(q_1 + q_2 + q_3) - m_{\text{t}}gL_{\text{s}} \sin(q_1) \\ & - m_{\text{t}}gL_{\text{Gt}} \sin(q_1 + q_2) - m_{\text{s}}gL_{\text{Gs}} \sin(q_1) \end{aligned} \tag{2.96}$$

$$\begin{aligned} G_2(\boldsymbol{q}) = & -m_{\text{ub}}gL_{\text{Gub}} \sin(q_1 + q_2 + q_3) - m_{\text{t}}gL_{\text{Gt}} \sin(q_1 + q_2) \\ & - m_{\text{ub}}gL_{\text{t}} \sin(q_1 + q_2) \end{aligned} \tag{2.97}$$

$$G_3(\boldsymbol{q}) = -m_{\text{ub}}gL_{\text{Gub}} \sin(q_1 + q_2 + q_3) \tag{2.98}$$

Using the same method, we can obtain the dynamic model of swing leg. Because the book's length is limited, we will not show here, and in the next chapters, these two models will be the research object.

2.4.4 Properties of the Model

For dynamic equation as shown by (2.74), we can prove that it meets the following properties [1]:

1. Positive definiteness

For arbitrary $\boldsymbol{q}$, matrix $\boldsymbol{H}(\boldsymbol{q})$ is positive definite.

2. The boundedness

For all the $\boldsymbol{q}, \dot{\boldsymbol{q}}$, matrix function $\boldsymbol{H}(\boldsymbol{q})$ and $\boldsymbol{C}(\boldsymbol{q}, \dot{\boldsymbol{q}})$ is uniformly bounded, that is, there is positive number λ_m, λ_n and positive definite function. Makes

$$0 \leq \lambda_m \boldsymbol{I} \leq \boldsymbol{H}(\boldsymbol{q}) \leq \lambda_n \boldsymbol{I} \tag{2.99}$$

$$\boldsymbol{C}^{\mathrm{T}}(\boldsymbol{q}, \dot{\boldsymbol{q}})\boldsymbol{C}(\boldsymbol{q}, \dot{\boldsymbol{q}}) \leq \boldsymbol{\eta}(\dot{\boldsymbol{q}})\boldsymbol{I} \tag{2.100}$$

3. The skew symmetry

For arbitrary $\boldsymbol{q}, \dot{\boldsymbol{q}}$, matrix function $\dot{\boldsymbol{H}}(\boldsymbol{q}) - 2 \cdot \boldsymbol{C}(\boldsymbol{q}, \dot{\boldsymbol{q}})$ is skew symmetric. Namely, for an arbitrary vector quantity ξ, there is

$$\xi^{\mathrm{T}}[\dot{\boldsymbol{H}}(\boldsymbol{q}) - 2 \cdot \boldsymbol{C}(\boldsymbol{q}, \dot{\boldsymbol{q}})]\xi = 0 \tag{2.101}$$

4. Linear characteristics

The exoskeleton suit's mathematical model is linear for physical parameter.

Namely, if we express the constant coefficient of the matrix function $\boldsymbol{H}, \boldsymbol{C}, \boldsymbol{G}$ as a vector quantity $\boldsymbol{\theta}$, we can define the proper matrix $\boldsymbol{\Phi}(\boldsymbol{q}, \dot{\boldsymbol{q}}, \boldsymbol{v}, \boldsymbol{a})$ and makes

$$\boldsymbol{H}(\boldsymbol{q}) \cdot \boldsymbol{a} + \boldsymbol{C}(\boldsymbol{q}, \dot{\boldsymbol{q}}) \cdot \boldsymbol{v} + \boldsymbol{G}(\boldsymbol{q}) = \boldsymbol{\Phi}(\boldsymbol{q}, \dot{\boldsymbol{q}}, \boldsymbol{v}, \boldsymbol{a})\boldsymbol{\theta} \tag{2.102}$$

hold.

In the equation, $\boldsymbol{v}$ is the velocity vector and $\boldsymbol{a}$ is the acceleration vector.

2.5 Human-Machine Interaction Model

In fact, the exoskeleton suit separated from the human is meaningless, and exoskeleton suit is a human-machine system. In the human-machine system, the interaction between human and the machine has been the important part of human-machine relationship. One of the important research directions of human-machine system is to realize the convenient communication between human and the machine. Taken together, the human-machine interactive way can be roughly divided into the following kinds: [13–15]

(1) In the traditional sense, operating handle, console, hand (foot) switch, and some others all are the important communication bridge between people and machines.
(2) With the advent of the computer, the interaction between human and the machine has obtained unprecedented development. To realize the interaction

between human and the machine by computer is considered to be the most ideal human-machine interactive way, and the interaction between human and the computer is realized through a user interface.

(3) For the multi-model interaction, the channels cover all kinds of communication methods of users expressing their intents, performing action, or perceiving feedback information, such as language, eye contact, facial expressions, lip movement, hand movement, gestures, head movement, body posture, touch, smell, or taste, and the computer user interface adopting this way is called the "multi-model user interface."

However, after the analysis, we have found that to realize the human-machine interaction of the exoskeleton suit system through the user interface is not appropriate. It is not possible for operators (the human) to wear the exoskeleton suit to hold the keyboard, mouse, and monitor to operate and send control instructions to exoskeleton suit constantly. The commonly used visual interaction needs several fixed cameras to capture the human body's moving posture [14], which is also not appropriate for the exoskeleton suit, because there is no fixed space to install these cameras. Others such as voice, handwriting, posture, sight, and expression are more inappropriate.

The most appropriate interactive way for exoskeleton suit should be touch, which is the most direct and natural interactive way between human and the machine. The touch has four basic functions: touch sense, sliding sense, force sense, and pressure sense. Among them, the force sense is three-dimensional force and three-dimensional torque and is touch's most complex, most comprehensive, and most widely used perception form [16, 17]. If we make the exoskeleton suit have force sense perception ability, then the exoskeleton suit can perceive the human body's movement intention by contacting with the human body naturally to generate the corresponding control command and control the exoskeleton suit to track human movement. The exoskeleton suit's sensitivity amplification control method and the force control method that will be introduced from Chap. 3–7 both require the natural contact of the human and exoskeleton suit to control exoskeleton suit's movement. In the sensitivity amplification control, relying on the force/torque the human body acting on the exoskeleton suit when they are contacting to change the motion state of the exoskeleton suit, and after the sensor installed in the exoskeleton suit have received the change of its motion state, the control signal will be formed to control the joint rotation of the exoskeleton suit, making the exoskeleton suit track human movement. When their movement is consistent, the contact force/torque the human acting on the exoskeleton suit becomes zero, and if keeping the cycle continuously, even the force and torque acted by the human is much lesser, they can drive the exoskeleton suit to move. In the exoskeleton suit's control method, using the multi-axis force/torque sensors to measure the contact force exerted by the human at the end of the exoskeleton suit, and using this information to generate the control signal of the exoskeleton suit to control the exoskeleton suit tracking human movement, then the same effect as that of the sensitivity amplification control method will be realized.

Then, how is the acting force between human and the machine generated? In fact, human and the exoskeleton suit are coupled together by certain links, such as waistcoat, waistband, thigh straps, shank straps, and shoes. When there is error in motion trajectory between human and the exoskeleton suit, the mutual acting force between them will be produced. This force is naturally produced in the process of movement, and there is no need to know the specific mathematical relationship between the error and the force. But if we want to simulate the exoskeleton suit system in computer, the mathematical relationship between them must be determined, namely establishing the mathematical model of human-machine acting force. In general, if we consider the person as an object or environment, then the property of interaction between the exoskeleton suit and human is the same as that of interaction between the robot and environment. The interactive force between the robot and environment can be thought as rigidity, the linear combination of spring and damping [18, 19], as shown in Eq. (2.103)

$$\boldsymbol{f}_{\mathrm{e}} = \boldsymbol{K}_{\mathrm{Pf}}(\boldsymbol{x} - \boldsymbol{x}_{\mathrm{e}}) + \boldsymbol{K}_{\mathrm{Df}}\dot{\boldsymbol{x}} + \boldsymbol{K}_{\mathrm{Mf}}\ddot{\boldsymbol{x}} \tag{2.103}$$

In the equation, $\boldsymbol{f}_{\mathrm{e}}$ expresses the force vector of robots acting on the environment; $\boldsymbol{x}, \dot{\boldsymbol{x}}, \ddot{\boldsymbol{x}}$ expresses position vector, velocity vector, and acceleration vector of the end of robots in the operating space; $\boldsymbol{x}_{\mathrm{e}}$ expresses the position vector of environment; $\boldsymbol{K}_{\mathrm{Pf}}$, $\boldsymbol{K}_{\mathrm{Df}}$, and $\boldsymbol{K}_{\mathrm{Mf}}$ express the spring coefficient matrix (the rigidity coefficient matrix), damping coefficient matrix, and inertia coefficient matrix, respectively. And some use nonlinear model to describe the environment's dynamic condition [18], as shown in Eqs. (2.104) and (2.105)

$$\boldsymbol{f}_{\mathrm{e}} = \boldsymbol{K}_{\mathrm{Pf}}(\boldsymbol{x} - \boldsymbol{x}_{\mathrm{e}})^{a} + \boldsymbol{K}_{\mathrm{Df}}\dot{\boldsymbol{x}}^{b} \tag{2.104}$$

$$\boldsymbol{f}_{\mathrm{e}} = \boldsymbol{K}_{\mathrm{Pf}}(\boldsymbol{x} - \boldsymbol{x}_{\mathrm{e}})^{c} + \boldsymbol{K}_{\mathrm{Df}}(\boldsymbol{x} - \boldsymbol{x}_{\mathrm{e}})^{c}\dot{\boldsymbol{x}} \tag{2.105}$$

For the sake of simplicity, this book does not use the nonlinear model, but is based on the linear model to research. And the environment described by Eq. (2.103) is static, while in the exoskeleton suit system, the human as the environment is kinetic. Then, assume the value of force vector of human acting on exoskeleton suit is $\boldsymbol{f}$ (in contrast to the $\boldsymbol{f}_{\mathrm{e}}$ sign); $\boldsymbol{x}_{\mathrm{h}}, \dot{\boldsymbol{x}}_{\mathrm{h}}, \ddot{\boldsymbol{x}}_{\mathrm{h}}$ express the information of human movement; $\boldsymbol{x}_{\mathrm{e}}, \dot{\boldsymbol{x}}_{\mathrm{e}}, \ddot{\boldsymbol{x}}_{\mathrm{e}}$ express the information of the exoskeleton suit movement; then, the $\boldsymbol{f}$ can be expressed as

$$\boldsymbol{f} = \boldsymbol{K}_{\mathrm{Pf}}(\boldsymbol{x}_{\mathrm{h}} - \boldsymbol{x}_{\mathrm{e}}) + \boldsymbol{K}_{\mathrm{Df}}(\dot{\boldsymbol{x}}_{\mathrm{h}} - \dot{\boldsymbol{x}}_{\mathrm{e}}) + \boldsymbol{K}_{\mathrm{Mf}}(\ddot{\boldsymbol{x}}_{\mathrm{h}} - \ddot{\boldsymbol{x}}_{\mathrm{e}}) \tag{2.106}$$

When taking no account of inertia coefficient, Eq. (2.106) can be simplified to a spring damping model

$$\boldsymbol{f} = \boldsymbol{K}_{\mathrm{Pf}}(\boldsymbol{x}_{\mathrm{h}} - \boldsymbol{x}_{\mathrm{e}}) + \boldsymbol{K}_{\mathrm{Df}}(\dot{\boldsymbol{x}}_{\mathrm{h}} - \dot{\boldsymbol{x}}_{\mathrm{e}}) \tag{2.107}$$

If we do not consider inertia coefficient and damping coefficient, Eq. (2.1
be simplified as spring model [20]

$$f = K_{\mathrm{Pf}}(x_{\mathrm{h}} - x_{\mathrm{e}}) \tag{2.108}$$

The force human acting on the exoskeleton suit and the force exoskeleton suit acting on human are the mutual reaction force, which can be called human-machine acting force. The human-machine acting force in this book refers to the force human made on the exoskeleton suit. In the later chapters, according to the human-machine interaction model (environment model) described by Eqs. (2.106)–(2.108), we will simulate the human-machine interaction in the computer to complete the simulation of various control methods.

References

1. Shen Tielong. The Basis of Robot's Robost Control [M]. Beijing: Tsinghua University Press, 2000.
2. Wang Yaonan. Robot Intelligent Control Engineering [M]. Beijing: Science Press, 2004.
3. Chen Ken, Yang Xiangdong, Yang Dongchao. Robot Technology and Application [M]. Beijing: Tsinghua University Press, 2006.
4. Xie Cunxi, Zhang Tie. Robot Technology and Applications [M]. Beijing: China Machine Press, 2005.
5. Kajita Hojo (Author). Guan Yisheng (Translator). Humanoid Robot [M]. Beijing: Tsinghua University Press, 2007.
6. Chen Zhanfu. The Modeling and Control Method Research of Lower Extremity Exoskeleton Suit [D]. Naval Aeronautical Engineering Institute, Master's Thesis, Yantai: 2008.
7. Racine, Jean-Louis Charles. Control of a Lower Extrmity Exoskeleton for Human Performance Amplification [D]. University of California, Ph. D, Berkeley: 2003.
8. Yang Qing. Real Time Motion Planning Method Research of Humanoid Robot [D]. National University of Defense Technology, Master's Thesis, Changsha: 2005.
9. Chen Lesheng. Geocentric Coordinate System of Kane Method [J]. Robot, 1994, 16 (3): 176–180.
10. Liu Minjie, Tian Yongtao, Li Congxin. Dynamics of Parallel Manipulator Using Sbu-Structure Kane Method [J]. Journal of Shanghai Jiaotong University, 2001, 35 (7): 1032–1035.
11. Yang Yuanming, Dong Qiuquan. KANE Equation of Forms of Energy. [J]. Journal of Shanxi Institute of Technology, 1997, 13 (2): 91–94.
12. Zheng Xiuyuan. Modern Sports Biomechanics [M]. Beijing: National Defence Industry Press, 2002.
13. Chen Ying, Yang Canjun. The Human-Machine Intelligent System [M]. Hangzhou: Zhejiang University Press, 2006.
14. Dong Shihai. Progress and Challenge of Human-Computer Interaction [J]. Journal of Computer-Aided Design & Computer, 2004, 16 (1): 1–13.
15. Ma Wenjuan, Fang Zhigang. Human-Computer Interaction Style and its Development Trends [J].
16. Yin Yuehong, Wei Zhongxin, Huang Xiaoxi. Force Sensing and Force Control Technology for Intelligent Machine System [M]. Beijing: National Defence Industry Press, 2002.

17. Yin Yuehong, Zhu Jianying. Force Sense and Control for Intelligent Machine: An Overview [J]. Chinese Journal of Aeronautics, 1999, 20 (1): 1–7.
18. Heng, Wang, K. H. Low, Michael Yu Wang. Reference Trajectory Generation for Force Tracking Impedance Control by Using Neural Network-based Environment Estimation [C]. 2006 IEEE Conference on Robotics Automation and Mechatronics, Bangkok, Thailand, 2006: 1–6.
19. Luo Yangyu, Wang Dangxiao, Zhang Yuru. Modeling and Analysis of a Force Interactive System [J]. Journal of Beijing University of Aeronautics and Astronautics, 2004, 30 (6): 539–542.
20. Murat, Cenk Cavusoglu, Joseph Yan, S. Shankar Sastry. A Hybrid System Approach to Contact Stability and Force Control in Robotic Manipulators [C]. In Proceedings of the 12th IEEE International Symposium on Intelligent Control (ISIC'97), Istanbul, Turkey, 1997: 143–148.

Chapter 3
Sensitivity Amplification Control of Exoskeleton Suit

3.1 Introduction

The exoskeleton suit combines the advantages of the human intelligence provided by the operator and of the mechanical energy provided by the exoskeleton suit. That is to say, the operator provides the intelligent control system for the exoskeleton suit, and the exoskeleton suit provides most of the energy needed for walking [1, 2]. In general, by measuring the motion locus of the operator, controlling exoskeleton suit following the human body's movement is the most direct idea. However, for the current technology, to measure the human body's motion locus only by the way of measuring information of human body joint's movement, this requires to install complex joint angle sensor, angular velocity sensor, and even angular acceleration sensor in the human body. These sensors are installed on the operator in various forms, which will cause the operator's comfort to be reduced greatly when using the exoskeleton suit. Moreover, the operator needs to install the sensors on the human body first whenever using the exoskeleton suit, and when the operator does not use the exoskeleton suit, the sensors also need to be removed, which will prolong the time for putting on and taking off. It can be endured for common civil use, but for soldiers, when facing danger, this will hinder the movement of soldiers, affecting the combat effectiveness.

So, are there such methods that not measuring the human body's movement, or not installing sensors on human body, and can the exoskeleton suit be controlled to follow the human movement? American BLEEX exoskeleton suit just uses this method. For the BLEEX exoskeleton suit control scheme, it does not need to measure the operator' body or the human-machine contact place (e.g., there is no pressure sensor between human and the machine) directly; on the contrary, just based on the measurement of the exoskeleton suit, the controller estimates how the exoskeleton suit will move and so make the operator receive very small force. The control scheme has never applied to any robot system. The scheme defines the ratio between the rotation angle of exoskeleton suit and the generalized force exerted by

Z. Yang et al., *Force Control Theory and Method of Human Load Carrying Exoskeleton Suit*, DOI 10.1007/978-3-662-54144-9_3

the human (refers to the force and torque acting on the object) as the sensitivity coefficient, and the purpose of this control method is to maximize the sensitivity coefficient, so it is called sensitivity amplification control (SAC) [1, 3, 4]. Its physical meaning can be explained as follows: If the exoskeleton suit rotates in the same angle, the greater the sensitivity coefficient is, the smaller the generalized force exerted by the human is. The following will analyze the sensitivity control method deeply.

3.2 Sensitivity Amplification Control

First of all, we illustrate the definition of sensitivity and the principle of SAC with an example of single degree of freedom. Figure 3.1 shows a human exoskeleton suit system schematic diagram of single degree of freedom and human-machine integration. Among them, T represents the total torque exerted on the exoskeleton suit's hip joint, q represents the hip joint angle of the exoskeleton suit, and m and l represent the mass and length of the exoskeleton suit, respectively. The diagram can be described as a scenario: The human's one leg stretches straightly (not flexion, the knee joint); the upper part of the body remains still; the exoskeleton suit's knee joint does not move and is limited to zero (nor flexion either); and the exoskeleton suit of single degree of freedom has a rotational degree of freedom with the human body in the place of hip joint and is tied together with the human body in the ankle. In this state, the leg can carry the exoskeleton suit to do rotational motion of single degree of freedom revolving around the hip joint. To describe the sensitivity amplification process clearly, the above scenario can be divided into two cases for discussion: One is that the all motoring torque of the exoskeleton suit is exerted by the human, and that is to say, there is no drive motor in the exoskeleton suit's hip joint (or drive motor does not work), and the other is that installing the angle sensor, angular velocity sensor, and angular acceleration sensor on the exoskeleton suit, and

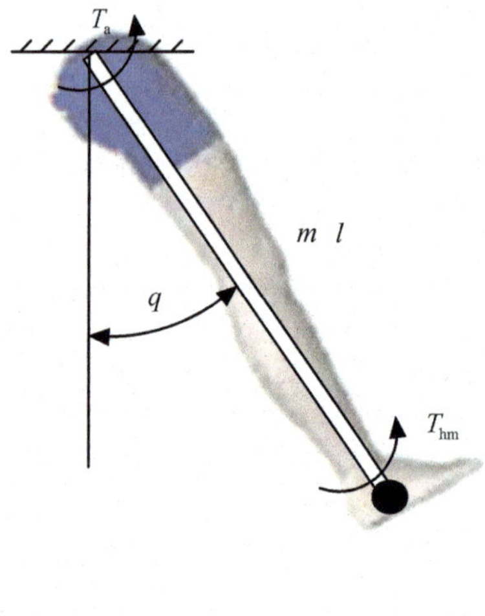

Fig. 3.1 Schematic diagram of the exoskeleton suit of single degree of freedom

installing drive motor on the hip joint, the exoskeleton suit's motoring torque is composed of the torque exerted by the human and the torque exerted by the exoskeleton suit.

3.2.1 1-DOF Exoskeleton Suit Without Actuation

The system block diagram of exoskeleton suit's hip joints without actuation is as shown in Fig. 3.2. Among them, the G represents the exoskeleton suit's system dynamic model, and the G_h represents the human-machine interaction model. The q_h represents the exoskeleton suit's desired trajectory, and q represents the exoskeleton suit's output trajectory. T_{hm} represents the torque exerted by the operator on the exoskeleton suit. T represents the sum of external torque exerted on the exoskeleton suit except gravity torque, and because the exoskeleton suit's hip joint is without actuation,

$$T = T_{hm} \tag{3.1}$$

Assume the system dynamic model is

$$q = G(T) \tag{3.2}$$

The inverse model is

$$T = G'(q) \tag{3.3}$$

Then by the equations (3.1) and (3.3), we can get

$$T_{hm} = G'(q) \tag{3.4}$$

If the system is linear, the equation (3.4) can be written as

$$T_{hm} = G' \cdot q \tag{3.5}$$

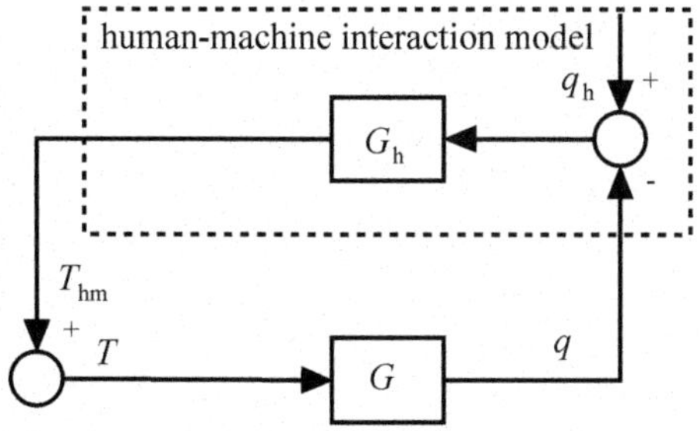

Fig. 3.2 System block diagram of exoskeleton suit without actuation

Define the sensitivity coefficient as

$$S = \frac{q}{T_{\rm hm}} \tag{3.6}$$

By the equation (3.5), we can get

$$S = \frac{1}{G'}. \tag{3.7}$$

3.2.2 1-DOF Exoskeleton Suit with Actuation

If we have added a drive motor to the exoskeleton suit's hip joint, and the motor's control signal is calculated by the controller according to the information of the exoskeleton suit itself, as shown in Fig. 3.3, design controller as

$$T_{\rm a} = (1 - \alpha^{-1})G'(q) \tag{3.8}$$

Among them, α is an amplification coefficient greater than 1. So the total torque exerted on the exoskeleton suit's hip joint is

$$T = T_{\rm hm} + T_{\rm a} \tag{3.9}$$

Then, by Eqs. (3.3) and (3.10), we can get the torque exerted by the operator as

$$T_{\rm hm} = \alpha^{-1}G'(q) \tag{3.10}$$

Similarly, if the system is linear, Eq. (3.11) can be rewritten as

$$T_{\rm hm} = \alpha^{-1} \cdot G' \cdot q \tag{3.11}$$

Then, by the definition of Eq. (3.7), we can get the new sensitivity coefficient

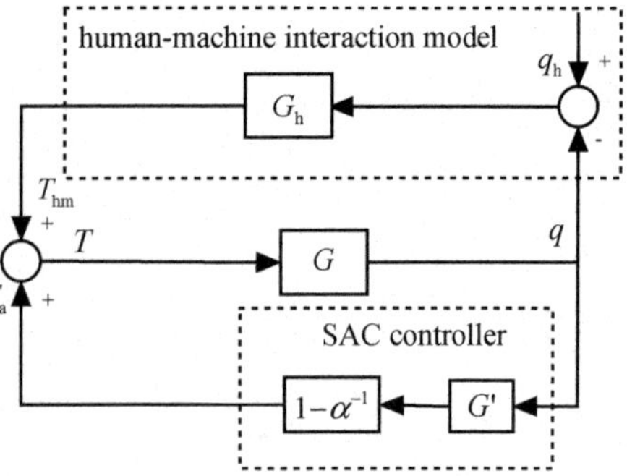

Fig. 3.3 System block diagram of exoskeleton suit with SAC controller

$$S_{\text{new}} = \alpha \cdot \frac{1}{G'} \tag{3.12}$$

By Eqs. (3.13) and (3.8), we can know

$$S_{\text{new}} = \alpha \cdot S \tag{3.13}$$

Therefore, if we choose $\alpha = 10$, after comparing Eq. (3.12) with Eq. (3.6), we can know that the torque exerted by the operator is only about one-tenth of the original, becoming very small. In Eq. (3.14), the sensitivity coefficient becomes ten times of the original, very large. This suggests that the SAC control can effectively reduce the torque of the operator exerting on exoskeleton suit system, reducing the operator's energy expenditure.

From Fig. 3.3, we can see that if we consider the human's acting force as a disturbance for the exoskeleton suit, then the exoskeleton suit system is a closed-loop control system; in classical control and modern control theory, the requirement for the controller is to minimize the sensitivity coefficient of whole closed-loop system for the external force and torque disturbance, which needs the human to exert great force to drive the exoskeleton suit, and even the human cannot drive the exoskeleton suit at all. And our basic requirements for the exoskeleton suit control are as follows: The exoskeleton suit needs to follow the human body's movement quickly and no laggingly, consciously, or unconsciously. To respond to all the force and torque on the exoskeleton suit, especially the force exerted by the human body, it requires very high sensitivity. But we have found that maximizing the system's sensitivity for external force and torque leads to loss of system's robustness [5]. Therefore, to improve the system's sensitivity needs compromising thinking.

To use this method, we also need to make two practical problems clear: The first, the exoskeleton suit with high sensitivity for the external force and torque will respond to other external force not made by the human. For example, if someone has pushed the exoskeleton suit with high sensitivity, it will move as driven by the operator. Although responding to other force makes the system itself not stable, which sounds like a serious problem, due to the instability of exoskeleton suit, the exoskeleton suit and the operator are not synchronized, so the operator actually receives the unexpected force from the exoskeleton suit and the operator will struggle to avoid unwanted movement, resulting in the reactive force to the exoskeleton suit, and SAC controller will keep stable under the function of the force. When receiving external force, the key to maintaining the exoskeleton suit stable and to preventing it from falling is depending on whether the operator can move quickly (moving backward or sideward) to form a stable state for himself (herself) and the exoskeleton suit. Therefore, the needing of a very wide control bandwidth makes the exoskeleton suit be able to respond to the initiative movement or noninitiative movement of human body.

The second is as follows: The system with high sensitivity for external force and torque is not good for the robustness of the parameter change; therefore, the precision of the system performance will be proportional to that of the exoskeleton suit's dynamic model. Although this is a very serious disadvantage, we can see that the increase of manpower caused by the model's inaccuracy is smaller than the increase of manpower caused by no SAC control, so it is feasible to implement this control method by improving the precision of system model as much as possible.

3.2.3 Sensitivity Amplification Control of Nonlinear System

In Eqs. (3.6) and (3.12), the systems both are assumed to be linear, and the linear system can get the system's expression of sensitivity coefficient. However, the exoskeleton suit is a typical nonlinear system, so it cannot get explicit representation of sensitivity coefficient. For example, for the system of single degree of freedom shown in Fig. 3.1, when the gravity term is not considered, the system is linear, and when the gravity term is considered, the system is nonlinear. What is shown in Eq. (3.3) is the nonlinear model of the system

$$T = J\ddot{q} + \frac{ml}{2}\sin q \tag{3.14}$$

Therefore, without adding actuator to the system, we can get

$$T_{\mathrm{hm}} = J\ddot{q} + \frac{ml}{2}\sin q \tag{3.15}$$

Obviously, we cannot get the explicit expression of the sensitivity coefficient on the basis of the definition Eq. (3.7) of sensitivity coefficient. But through the proper controller design, and according to the **weak or strong** manpower, we still can judge whether the sensitivity of the system has been increased or not. For example, if the controller is designed as

$$T_{\mathrm{a}} = (1 - \alpha^{-1})J\ddot{q} + \frac{ml}{2}\sin q \tag{3.16}$$

then under the function of the controller, the torque exerted by the human becomes

$$T_{\mathrm{hm}} = \alpha^{-1}J\ddot{q} \tag{3.17}$$

If we choose $\alpha = 10$, after comparing Eq. (3.18) with Eq. (3.16), it can be seen that the torque exerted by human decreases significantly and the sensitivity of system is magnified when the exoskeleton suit does the same movement locus.

For the system of multiple degrees of freedom, referring to dynamic Eq. (2.74), when considering the friction, the dynamic model of the system can be expressed as

$$\boldsymbol{H}(\boldsymbol{q})\ddot{q}+\boldsymbol{C}(\boldsymbol{q},\dot{q})\dot{q}+\boldsymbol{F}\dot{q}+\boldsymbol{G}(\boldsymbol{q})=\boldsymbol{T}_{\mathrm{a}}+\boldsymbol{T}_{\mathrm{hm}} \tag{3.18}$$

In the equation, $\boldsymbol{H}(\boldsymbol{q})$ is inertia matrix; $\boldsymbol{C}(\boldsymbol{q},\dot{\boldsymbol{q}})$ is Coriolis term; $\boldsymbol{F}$ represents movement friction coefficient matrix; $\boldsymbol{G}(\boldsymbol{q})$ is gravity term matrix; $\boldsymbol{T}_{\mathrm{a}}$ represents torque vector exerted by actuator; and $\boldsymbol{T}_{\mathrm{hm}}$ represents the human-machine torque vector, i.e., the torque exerted by the operator.

Obviously, if the actuator does not work, namely when $\boldsymbol{T}_{\mathrm{a}}=0$, there is

$$\boldsymbol{T}_{\mathrm{hm}}=\boldsymbol{H}(\boldsymbol{q})\ddot{q}+\boldsymbol{C}(\boldsymbol{q},\dot{q})\dot{q}+\boldsymbol{F}\dot{q}+\boldsymbol{G}(\boldsymbol{q}) \tag{3.19}$$

If the controller is designed as

$$\boldsymbol{T}_{\mathrm{a}}=(1-\alpha^{-1})[\boldsymbol{H}(\boldsymbol{q})\ddot{q}+\boldsymbol{C}(\boldsymbol{q},\dot{q})\dot{q}]+\boldsymbol{F}\dot{q}+\boldsymbol{G}(\boldsymbol{q}) \tag{3.20}$$

then under the function of the controller, torque exerted by the human becomes

$$\boldsymbol{T}_{\mathrm{hm}}=\alpha^{-1}[\boldsymbol{H}(\boldsymbol{q})\ddot{q}+\boldsymbol{C}(\boldsymbol{q},\dot{q})\dot{q}] \tag{3.21}$$

If we choose appropriate α, after comparing Eq. (3.22) with Eq. (3.20), it can be seen that the torque exerted by the human decreases significantly and the sensitivity of system is magnified when the exoskeleton suit does the same movement locus.

Because the nonlinear system is unable to get explicit expression of sensitivity coefficient. Therefore, to better verify the controller's control effect, define another two no factor variables (human factor ***HF*** and maximum human factor ***HFM***). And the human factor ***HF*** is defined as the ratio between the human-machine acting force and the resultant external force acting on the exoskeleton suit, namely

$$\boldsymbol{HF}=\frac{\boldsymbol{T}_{\mathrm{hm}}}{\boldsymbol{T}}=\frac{\boldsymbol{T}_{\mathrm{hm}}}{\boldsymbol{T}_{\mathrm{hm}}+\boldsymbol{T}_{\mathrm{a}}} \tag{3.22}$$

The largest human factor (***HFM***) is defined as the ratio between the maximum value of the absolute value of human-machine acting force in the movement process and the maximum value of the absolute value of the resultant external force, namely

$$\boldsymbol{HFM}=\frac{\max|\boldsymbol{T}_{\mathrm{hm}}|}{\max|\boldsymbol{T}|}=\frac{\max|\boldsymbol{T}_{\mathrm{hm}}|}{\max|\boldsymbol{T}_{\mathrm{hm}}+\boldsymbol{T}_{\mathrm{a}}|} \tag{3.23}$$

The ideal ***HF*** value should be near zero, representing human-machine acting force accounting for a very small proportion in the whole movement process of the exoskeleton suit. Namely, the acting force exerted by human is small, showing that the controller's effect is obvious. Ideal ***HFM*** also should be near zero, representing the maximum force exerted by human is also very small in the movement process. However, neither ***HF*** nor ***HFM*** can fully reflect the system's control effect, and the human-machine acting force curve is needed to be combined together to comprehensively survey the control effect.

3.2.4 Stability Analysis

Only when the exoskeleton suit tracks human body's motion trajectory effectively, the human-machine torque shown in Eqs. (3.11) or (3.16) can be obtained. So the above controller's stability analysis must be made. From the perceptual intuition, it is certainly stable for the human body to operate a machine without dynamics, as long as the required force or moment does not exceed the scope of human ability. For example, for the system as shown in Fig. 3.3, if the controller does not work, all control torque of the exoskeleton suit is exerted by human, and because the human is intelligent, the system is stable.

For the stability analysis of SAC control, the exact human-machine interaction model needs to be known, namely G_{h}, and the specific form of G_{h} is unknown, just known that it acts on human body muscle force of the exoskeleton suit. Human body muscle model is very complex, including nerve conduction, muscle contraction, and the processing of the central nervous system, and we will not study them here. From the point of view of control, we assume that G_{h} is an operator, whose input is the motion state of the human body and exoskeleton suit, and whose output is force or torque the human body acting on the exoskeleton suit. Apparently, the operator is a complex nonlinear operator with intelligence, being unable to find its precise mathematical expression. However, in order to simulate the human-machine interaction in computer, to simulate the generalized force human acting on the exoskeleton suit, and to make the design and stability analysis of exoskeleton suit's controller, the reasonable hypothesis model must be established. For the detailed analysis of the human-machine interaction model, please refer to Sect. 2.4 of this book. Here, we choose the spring damping model shown in Eq. (2.107) to make the stability analysis of the SAC control system.

Refer to the human-machine interaction model shown in Eq. (2.107) and convert it to the expression means of torque and joint movement information in joint space. Then, the spring damping model of human-machine interaction is

$$\boldsymbol{T}_{\mathrm{hm}} = \boldsymbol{K}_{\mathrm{Pf}}(\boldsymbol{q}_{\mathrm{h}} - \boldsymbol{q}) + \boldsymbol{K}_{\mathrm{Df}}(\dot{\boldsymbol{q}}_{\mathrm{h}} - \dot{\boldsymbol{q}}) \tag{3.24}$$

If the designed controller is as shown in Eq. (3.21), after substituting Eq. (3.25) into Eq. (3.22), we can get

$$\boldsymbol{K}_{\mathrm{Pf}}(\boldsymbol{q}_{\mathrm{h}} - \boldsymbol{q}) + \boldsymbol{K}_{\mathrm{Df}}(\dot{\boldsymbol{q}}_{\mathrm{h}} - \dot{\boldsymbol{q}}) = \alpha^{-1}[\boldsymbol{H}(\boldsymbol{q})\ddot{\boldsymbol{q}} + \boldsymbol{C}(\boldsymbol{q}, \dot{\boldsymbol{q}})\dot{\boldsymbol{q}}] \tag{3.25}$$

Let $\Delta \boldsymbol{q} = \boldsymbol{q}_{\mathrm{h}} - \boldsymbol{q}$, $\Delta \dot{\boldsymbol{q}} = -\dot{\boldsymbol{q}}$, $\Delta \ddot{\boldsymbol{q}} = -\ddot{\boldsymbol{q}}$, then Eq. (3.26) becomes

$$\boldsymbol{H}(\boldsymbol{q})\ddot{\boldsymbol{q}} = \alpha \boldsymbol{K}_{\mathrm{Pf}}\Delta \boldsymbol{q} + \alpha \boldsymbol{K}_{\mathrm{Df}}\Delta \dot{\boldsymbol{q}} - \boldsymbol{C}(\boldsymbol{q}, \dot{\boldsymbol{q}})\dot{\boldsymbol{q}} \tag{3.26}$$

Introduce Lyapunov function

$$V = \frac{1}{2}\dot{\boldsymbol{q}}^{\mathrm{T}}\boldsymbol{H}(\boldsymbol{q})\dot{\boldsymbol{q}} + \frac{1}{2}\Delta\boldsymbol{q}^{\mathrm{T}}\boldsymbol{K}\Delta\boldsymbol{q} \tag{3.27}$$

Then, $V > 0$, differentiating the V, we can obtain

$$\begin{aligned}\dot{V} &= \frac{1}{2}\dot{\boldsymbol{q}}^{\mathrm{T}}\dot{\boldsymbol{H}}(\boldsymbol{q})\dot{\boldsymbol{q}} + \dot{\boldsymbol{q}}^{\mathrm{T}}\boldsymbol{H}(\boldsymbol{q})\ddot{\boldsymbol{q}} + \Delta\dot{\boldsymbol{q}}^{\mathrm{T}}\boldsymbol{K}\Delta\boldsymbol{q} \\ &= \dot{\boldsymbol{q}}^{\mathrm{T}}\alpha\boldsymbol{K}_{\mathrm{Pf}}\Delta\boldsymbol{q} + \dot{\boldsymbol{q}}^{\mathrm{T}}\alpha\boldsymbol{K}_{\mathrm{Df}}\Delta\dot{\boldsymbol{q}} + \Delta\dot{\boldsymbol{q}}^{\mathrm{T}}\boldsymbol{K}\Delta\boldsymbol{q} \\ &= -\Delta\dot{\boldsymbol{q}}^{\mathrm{T}}\alpha\boldsymbol{K}_{\mathrm{Pf}}\Delta\boldsymbol{q} - \Delta\dot{\boldsymbol{q}}^{\mathrm{T}}\alpha\boldsymbol{K}_{\mathrm{Df}}\Delta\dot{\boldsymbol{q}} + \Delta\dot{\boldsymbol{q}}^{\mathrm{T}}\boldsymbol{K}\Delta\boldsymbol{q}\end{aligned} \tag{3.28}$$

Let $\boldsymbol{K} = \alpha\boldsymbol{K}_{\mathrm{Pf}}$, then

$$\dot{V} = -\Delta\dot{\boldsymbol{q}}^{\mathrm{T}}\alpha\boldsymbol{K}_{\mathrm{Df}}\Delta\dot{\boldsymbol{q}} \tag{3.29}$$

Therefore, as long as the chosen matrix is positive definite, then, the $\dot{V}$ is not positive, so the system is asymptotically stable in the balance point.

3.2.5 *Simulation Analysis*

3.2.5.1 The Simulation of Single Degree of Freedom

According to the model shown in Fig. 3.1, use the MATLAB software to simulate. Considering the existence of gravity term in the simulation, then, the system is nonlinear, so we cannot contrast sensitivity coefficient directly, but can observe the control effect by the magnitude of human-machine moment, so the simulation can be done in three cases:

(1) Human's legs drive the exoskeleton suit's movement, and the exoskeleton suit has no other added actuator;
(2) Adding SAC controller, human's legs together with actuator drive the exoskeleton suit's movement and design the SAC controller as

$$T_{\mathrm{a}} = (1 - \alpha^{-1})\left(J\ddot{q} + \frac{ml}{2}\sin q\right) \tag{3.30}$$

(3) Adding SAC controller, human's legs together with actuator drive the exoskeleton suit's movement and design the SAC controller as shown in Eq. (3.17). We choose $\alpha = 8$ in the cases (2) and (3).

Assuming that the initial angle of human's legs is 5°, the angular acceleration of legs' movement does sinusoidal motion in the amplitude of $2°/s^2$ and in the frequency of $1/\pi$, and the angular velocity and angle are the integral value of angular acceleration and angular velocity, respectively. The simulation results are as shown in Figs. 3.4, 3.5, 3.6, 3.7, and 3.8.

Among them, Fig. 3.4 shows the exoskeleton suit angle's tracking curve for human body angle. Because the three cases of angle tracking curve are very similar, here we only provide the angle tracking curve without actuator. From the figure, it can be seen that in the three cases, the condition of the exoskeleton suit's tracking human body angle is very good. Figure 3.5 shows the torque's contrast curve of the cases (1) and (2). In the case (1), all control torque of the exoskeleton suit is exerted by the human, so the torque required by the human is very large, as shown by the full line. In the case (2), the control torque of the exoskeleton suit contains two parts, namely the torque exerted by the human shown in chain dotted line and the torque shown in dotted line exerted by the actuator. From the contrast in the figure, we can know after being added to the SAC controller, the actuator has exerted most

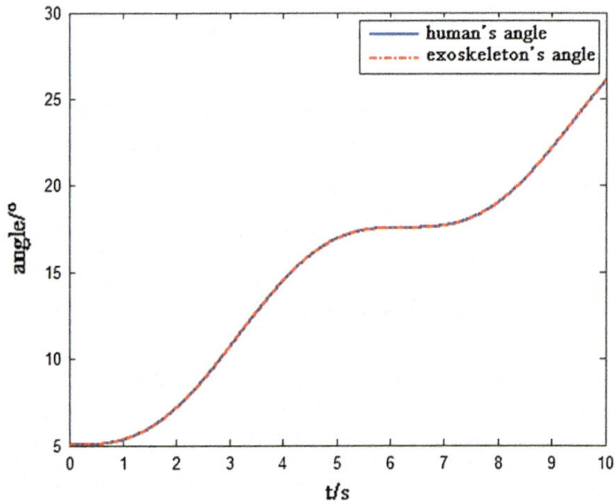

Fig. 3.4 Angle tracking *curve*

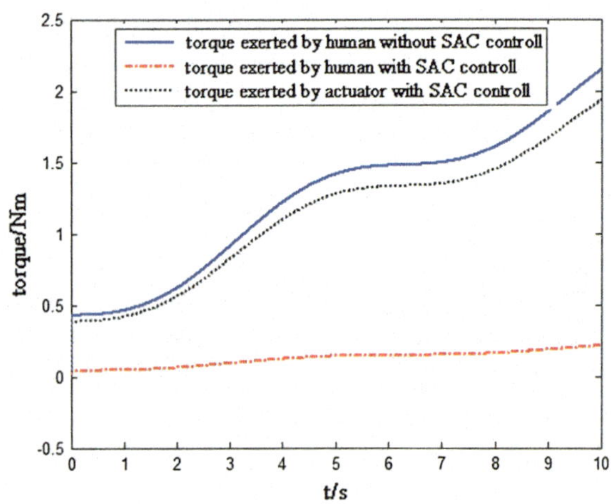

Fig. 3.5 Correlation curve of the torque in case (1) and case (2)

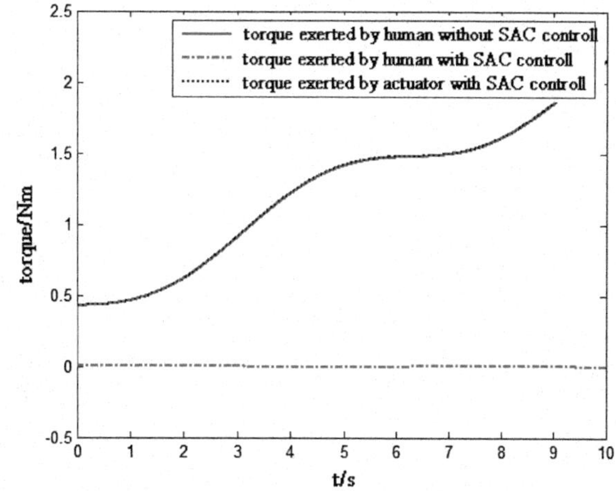

Fig. 3.6 Correlation curve of the torque in case (1) and case (3)

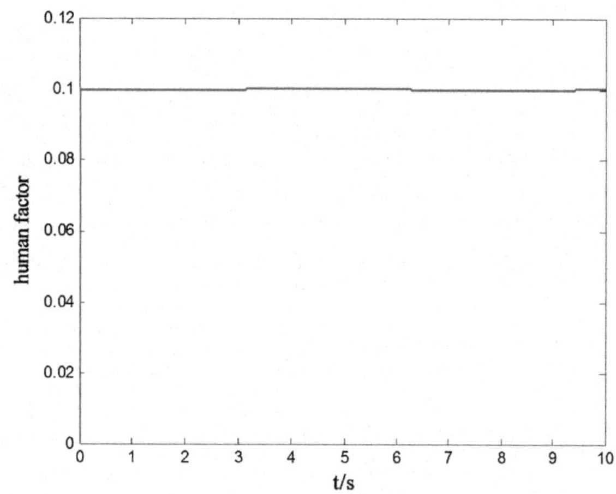

Fig. 3.7 Human factor in case (2)

of the control torque, and the torque exerted by the human becomes very small, which shows that SAC controller not only tracks the human body angle but also reduces the torque exerted by the human, thereby reducing the human's energy consumption.

Figure 3.6 shows the correlation curve of torque between the case (1) and case (3).

From the figure, we can know that compared with Fig. 3.5, the torque exerted by the human somewhat decreases. The difference between the case (3) and case (2) is that the controller designed in case (3) completely offsets the effect of gravity term, so the torque exerted by the human is smaller than the torque exerted by the human in case (2). The simulation results illustrate that the SAC control method can guarantee the exoskeleton suit's tracking human body movement and at the same time can reduce the energy consumption of the body.

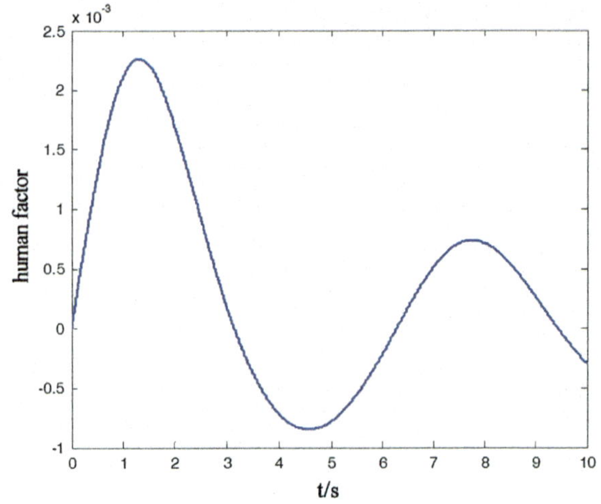

Fig. 3.8 Human factor in case (3)

Figure 3.7 shows the human factor curve in case (2). From Fig. 3.7, we can see that the human factor *HF* always remains in 0.1, indicating that the acting force exerted by the human accounts for only 10% of the movement process, and the corresponding maximum human factor *HFM* is 0.1. Figure 3.8 shows the human factor curve in case (3). From Fig. 3.7, we can see that the human factor is smaller, and in 10^{-3} orders of magnitude, the corresponding *HFM* is 0.001. Because the system is single degree of freedom, at this time, the *HF* and *HFM* both are a scalar.

3.2.5.2 The Simulation of Swing Leg

The example of single degree of freedom illustrates the working principle and effectiveness of SAC. Next, we will use the swing leg model of three degrees of freedom to make further illustration. The exoskeleton suit's swing leg model is as shown in Fig. 2.5b.

When simulating, assume the exoskeleton suit has the same geometry size and mass property as human; therefore, we can adopt the human body parameter of Winter as the parameter of the swing leg [6]. This group of parameters is experience parameters, so as long as the height *H* and weight *M* have been given, we can calculate the mass, length, center-of-gravity position, and rotational inertia of the other limbs' links. And for its calculation method, please refer to annex B.

At the same time, make use of Clinical Gait Analysis (CGA) data as the desired human movement signal. These data have given each joint's angle, angular velocity, and angular acceleration curve of human's lower limbs in the walking process, and please refer to appendix C for the data, from which choose the data in the swing phase to simulate. Because CGA data's step size is large and the data volume is insufficient, we can get other data by the interpolation method.

Use the above data and MATLAB to simulate. The controller is as shown in Eq. (3.21), and similarly, the simulation is carried out in two cases:

(1) For one case, the actuator is not added and all control torque of the exoskeleton suit is exerted by human and
(2) For the other case, the SAC controller has been added and the control torque of the exoskeleton suit is composed of the human's control torque and actuator's control torque.

The simulation results are as shown in Figs. 3.9, 3.10, 3.11, and 3.12. Figure 3.9 shows the joint angle tracking curve of the exoskeleton suit in the simulation of the case (1), and from the curve, we can see that the three joints of the exoskeleton suit all can track human movement well. Figure 3.10 shows the torque exerted on the joints of the exoskeleton suit by human in the simulation of case (2). Figure 3.11 shows the joint angle tracking curve of the exoskeleton suit in simulation of the case (2), and from the figure, we can see that the exoskeleton suit's angle tracking is more accurate. Figure 3.12 shows the contrast curve of the torque exerted by human

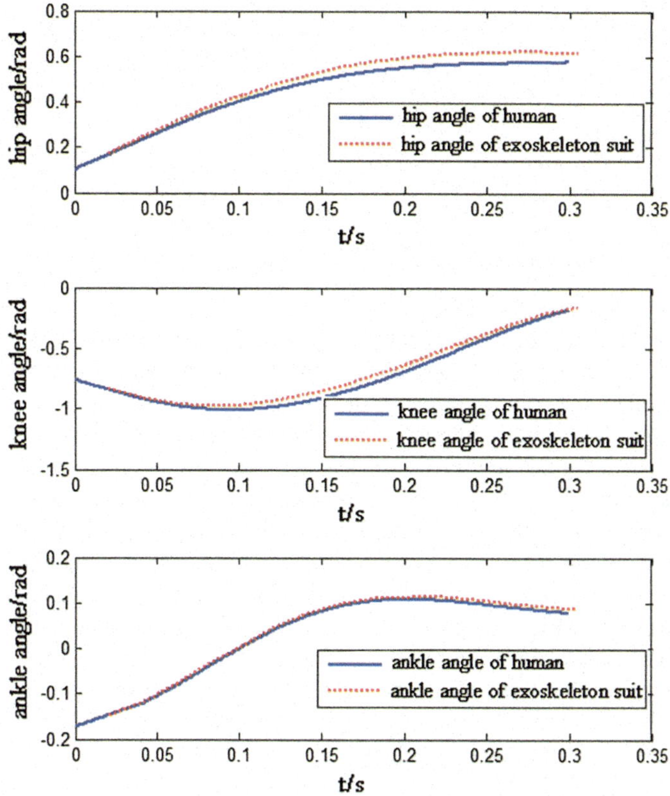

Fig. 3.9 Exoskeleton suit's angle tracking curve without adding SAC control

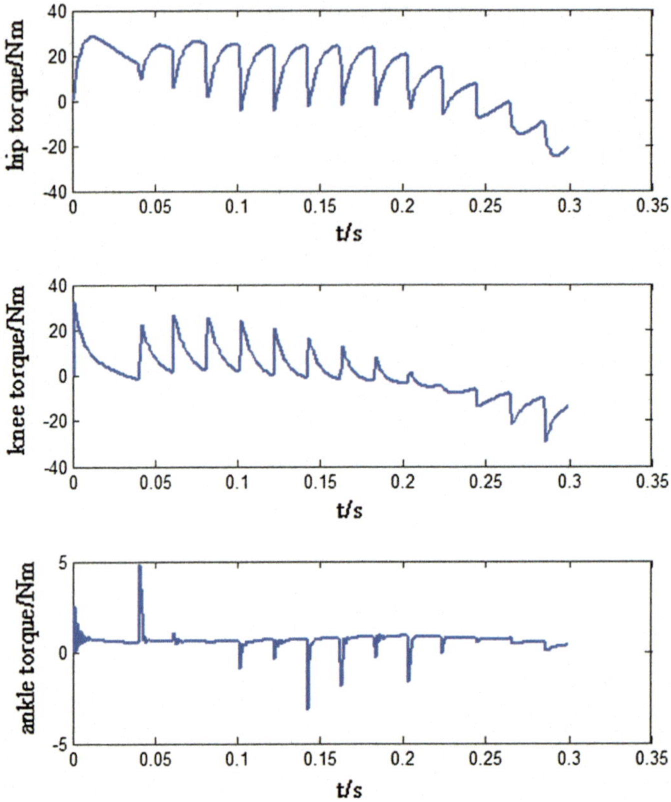

Fig. 3.10 Joint torque exerted by human without adding SAC control

and the torque exerted by the actuator in the simulation of case (2); comparing the two curves with the curve shown in Fig. 3.10, we can see that after SAC controller has been added, the control torque of the exoskeleton suit is mainly exerted by the actuator, and the human's torque is reduced, thus reducing human's energy consumption, so the goal of the control has been achieved. Figure 3.13 shows the human factor of joint space, and from the figure, we can see that most of the three joints' human factor ***HF*** is restricted to ± 0.3; the peak in the curve appears at the ZCP (zero-crossing point) of the resultant external force; at this time, the ***HF*** being close to, even more than 1, does not mean the force exerted by human is very large, so we need refer to Fig. 3.12 to analyze comprehensively to get the right result. The maximum human factor of the three joints is $\boldsymbol{HFM} = \begin{bmatrix} 0.1744 & 0.1326 & 0.1213 \end{bmatrix}$, respectively, and the three elements are, respectively, corresponding to the ankle, knee, and hip joint, and it is the same in the following sections. All this means that the operator can swing the exoskeleton suit very easily.

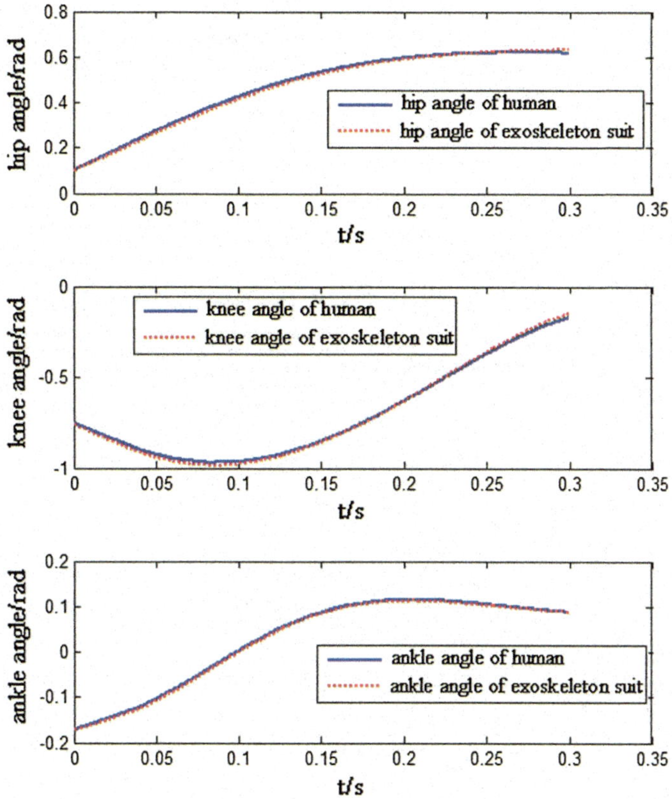

Fig. 3.11 Exoskeleton suit's joint angle tracking curve after the SAC control has been added

3.3 Sensitivity Amplification Control Based on Neural Network

From Eqs. (3.9) to (3.21), we can see that the SAC controller is heavily dependent on the system's dynamic model, so the system's accurate mathematical model must be established [5], but it is often difficult in the practical. And multilayer feedforward neural networks can approach arbitrary nonlinear mapping in arbitrary precision, bringing a new, unconventional tool for the modeling of complex system. We just need to know the input and output data reflecting the characteristics of the system when adopting the neural network to establish the system's dynamic model and through training obtain the network with a certain generalization ability to replace the system's mathematical model, which simplifies the modeling process. The block diagram of the neural network SAC controller is as shown in Fig. 3.14.

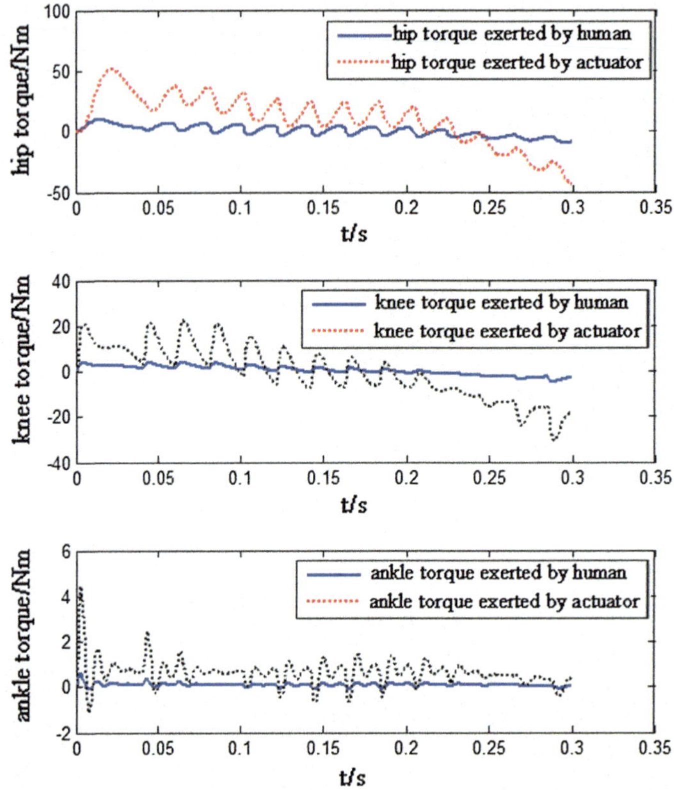

Fig. 3.12 Contrast curve of the torque exerted by human and the torque exerted by the actuator after the SAC control has been added

3.3.1 Stability Analysis

For the system's dynamic model as shown in Eq. (3.19), let

$$\boldsymbol{\Psi}(\boldsymbol{x}) = \boldsymbol{H}(\boldsymbol{q})\ddot{\boldsymbol{q}} + \boldsymbol{C}(\boldsymbol{q}, \dot{\boldsymbol{q}})\dot{\boldsymbol{q}} + \boldsymbol{F}\dot{\boldsymbol{q}} + \boldsymbol{G}(\boldsymbol{q}) \tag{3.31}$$

When there is not any control, namely when $\boldsymbol{T}_{\mathrm{a}} = 0$, the operator exerts all the control torque, namely

$$\boldsymbol{T}_{\mathrm{hm}} = \boldsymbol{\Psi}(\boldsymbol{x}) \tag{3.32}$$

Assume $\boldsymbol{\Psi}(\boldsymbol{x})$ is unknown, adopting the neural network to approach the $\boldsymbol{\Psi}(\boldsymbol{x})$, and according to the expression of $\boldsymbol{\Psi}(\boldsymbol{x})$, choose the network input

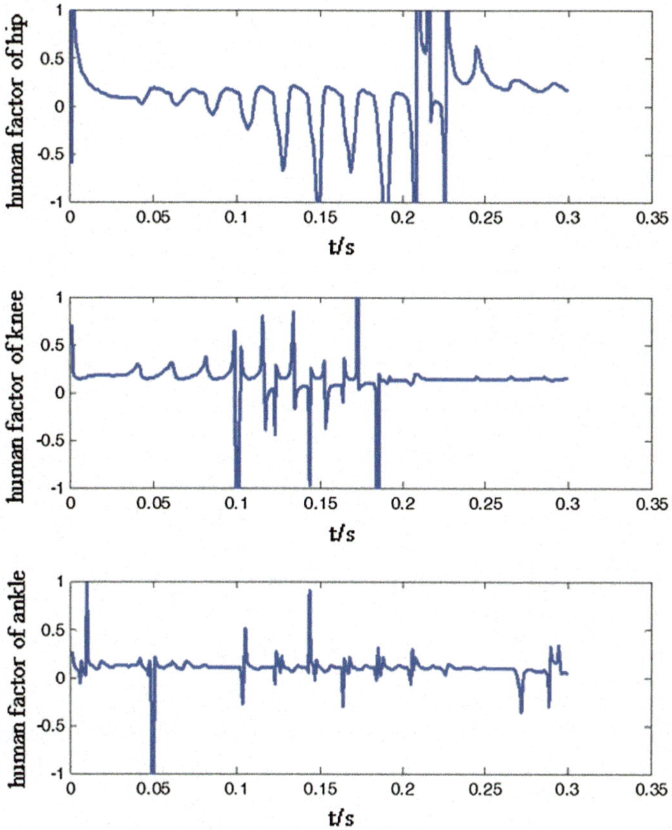

Fig. 3.13 Human factor of joint space

Fig. 3.14 System's block diagram of the exoskeleton suit with NNSAC controller

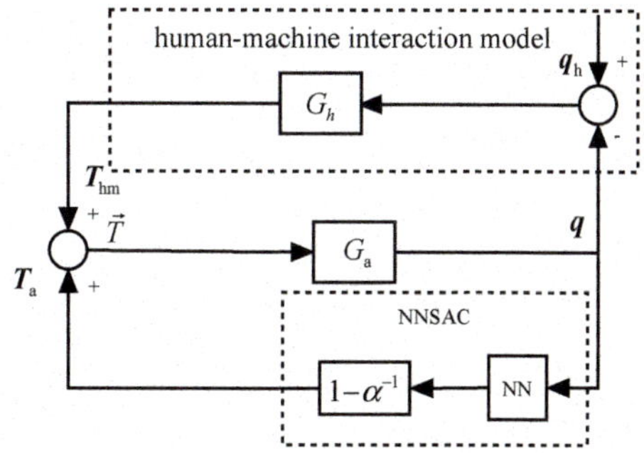

$$\boldsymbol{x} = [\boldsymbol{q} \quad \dot{\boldsymbol{q}} \quad \ddot{\boldsymbol{q}}]^T \tag{3.33}$$

The design control law is

$$\boldsymbol{T}_{\mathrm{a}} = (1 - \alpha^{-1})\hat{\boldsymbol{\Psi}}(\boldsymbol{x}) \tag{3.34}$$

In the equation, $\hat{\boldsymbol{\Psi}}(\boldsymbol{x})$ is the estimated value of the $\boldsymbol{\Psi}(\boldsymbol{x})$ by the neural network.

After substituting Eq. (3.32) and the control law Eq. (3.36) into Eq. (3.19), we will get

$$\boldsymbol{\Psi}(\boldsymbol{x}) = (1 - \alpha^{-1})\hat{\boldsymbol{\Psi}}(\boldsymbol{x}) + \boldsymbol{T}_{\mathrm{hm}} \tag{3.35}$$

So

$$\boldsymbol{T}_{\mathrm{hm}} = \tilde{\boldsymbol{\Psi}}(\boldsymbol{x}) + \alpha^{-1}\hat{\boldsymbol{\Psi}}(\boldsymbol{x}) \tag{3.36}$$

In the equation, $\tilde{\boldsymbol{\Psi}}(\boldsymbol{x}) = \boldsymbol{\Psi}(\boldsymbol{x}) - \hat{\boldsymbol{\Psi}}(\boldsymbol{x})$ is the approximation error of the neural network.

If the network's approximation error is zero, then

$$\boldsymbol{T}_{\mathrm{hm}} = \alpha^{-1}\boldsymbol{\Psi}(\boldsymbol{x}) \tag{3.37}$$

Then, when we choose $\alpha = 10$, the torque exerted by human is only one-tenth of the original. Even if the network's approximation error is not zero, as long as the error is small, then by comparing Eqs. (3.38) and (3.33), we can see that the torque exerted by human is still very small compared with the original.

In order to prove the system's stability, firstly, let

$$\boldsymbol{\Psi}'(\boldsymbol{x}) = \boldsymbol{H}(\boldsymbol{q})\ddot{\boldsymbol{q}} + \boldsymbol{C}(\boldsymbol{q}, \dot{\boldsymbol{q}})\dot{\boldsymbol{q}} + \boldsymbol{F}\dot{\boldsymbol{q}} \tag{3.38}$$

Then

$$\boldsymbol{\Psi}(\boldsymbol{x}) = \boldsymbol{\Psi}'(\boldsymbol{x}) - \frac{\alpha^{-1}\boldsymbol{G}(\boldsymbol{q})}{1 - \alpha^{-1}} + \frac{\boldsymbol{G}(\boldsymbol{q})}{1 - \alpha^{-1}} \tag{3.39}$$

Without loss of generality, we can assume

$$\hat{\boldsymbol{\Psi}}(\boldsymbol{x}) = \hat{\boldsymbol{\Psi}}'(\boldsymbol{x}) + \frac{\boldsymbol{G}(\boldsymbol{q})}{1 - \alpha^{-1}} \tag{3.40}$$

In the equation, $\hat{\boldsymbol{\Psi}}'(\boldsymbol{x})$ is the approximation value of $\boldsymbol{\Psi}'(\boldsymbol{x}) - \alpha^{-1}\boldsymbol{G}(\boldsymbol{q})/(1 - \alpha^{-1})$ by neural network, so

$$\tilde{\boldsymbol{\Psi}}(\boldsymbol{x}) = \boldsymbol{\Psi}(\boldsymbol{x}) - \hat{\boldsymbol{\Psi}}(\boldsymbol{x}) = \boldsymbol{\Psi}'(\boldsymbol{x}) - \frac{\alpha^{-1}\boldsymbol{G}(\boldsymbol{q})}{1 - \alpha^{-1}} - \hat{\boldsymbol{\Psi}}'(\boldsymbol{x}) \tag{3.41}$$

Namely, the output of neural network fully approximates the gravity term $\boldsymbol{G}(\boldsymbol{q})/(1-\alpha^{-1})$, so the approximation error is completely caused by the $\hat{\boldsymbol{\Psi}}'(\boldsymbol{x})$, and even though the output of neural network does not fully approximate the gravity term $\boldsymbol{G}(\boldsymbol{q})/(1-\alpha^{-1})$, it also can transfer the approximation error into the error of $\hat{\boldsymbol{\Psi}}'(\boldsymbol{x})$. So the control law Eq. (3.36) can be changed to

$$\boldsymbol{T}_{\mathrm{a}} = (1-\alpha^{-1})\hat{\boldsymbol{\Psi}}'(\boldsymbol{x}) + \boldsymbol{G}(\boldsymbol{q}) \tag{3.42}$$

After substituting Eqs. (3.42) and (3.32) into Eq. (3.19), we will get

$$\boldsymbol{\Psi}(\boldsymbol{x}) = (1-\alpha^{-1})\hat{\boldsymbol{\Psi}}'(\boldsymbol{x}) + \boldsymbol{G}(\boldsymbol{q}) + \boldsymbol{T}_{\mathrm{hm}} \tag{3.43}$$

Therefore, by Eqs. (3.32) and (3.40), we will get

$$\boldsymbol{\Psi}'(\boldsymbol{x}) = (1-\alpha^{-1})\hat{\boldsymbol{\Psi}}'(\boldsymbol{x}) + \boldsymbol{T}_{\mathrm{hm}} \tag{3.44}$$

Because $\alpha > 1$, there is positive definite function $\boldsymbol{k}(\boldsymbol{x})$, satisfying $\|\boldsymbol{k}(\boldsymbol{x})\| < 1$ and

$$(1-\alpha^{-1})\hat{\boldsymbol{\Psi}}'(\boldsymbol{x}) = \boldsymbol{k}(\boldsymbol{x})\boldsymbol{\Psi}'(\boldsymbol{x}) \tag{3.45}$$

So

$$\boldsymbol{\Psi}'(\boldsymbol{x}) = \boldsymbol{k}(\boldsymbol{x})\boldsymbol{\Psi}'(\boldsymbol{x}) + \boldsymbol{T}_{\mathrm{hm}} \tag{3.46}$$

After substituting the equation of human-machine interaction model (3.25) into Eq. (3.46), we can get

$$\boldsymbol{\Psi}'(\boldsymbol{x}) = [\boldsymbol{I}-\boldsymbol{k}(\boldsymbol{x})]^{-1}[\boldsymbol{K}_{\mathrm{Pf}}(\boldsymbol{q}_{\mathrm{h}}-\boldsymbol{q}) + \boldsymbol{K}_{\mathrm{Df}}(\dot{\boldsymbol{q}}_{\mathrm{h}}-\dot{\boldsymbol{q}})] \tag{3.47}$$

Let $\Delta\boldsymbol{q} = \boldsymbol{q}_{\mathrm{h}} - \boldsymbol{q}$, $\Delta\dot{\boldsymbol{q}} = -\dot{\boldsymbol{q}}$, $\Delta\ddot{\boldsymbol{q}} = -\ddot{\boldsymbol{q}}$ and substitute Eq. (3.40) into Eq. (3.47), then

$$\boldsymbol{H}(\boldsymbol{q})\ddot{\boldsymbol{q}} = [\boldsymbol{I}-\boldsymbol{k}(\boldsymbol{x})]^{-1}\boldsymbol{K}_{\mathrm{Pf}}\Delta\boldsymbol{q} + [\boldsymbol{I}-\boldsymbol{k}(\boldsymbol{x})]^{-1}\boldsymbol{K}_{\mathrm{Df}}\Delta\dot{\boldsymbol{q}} - \boldsymbol{C}(\boldsymbol{q},\dot{\boldsymbol{q}})\dot{\boldsymbol{q}} - \boldsymbol{F}\dot{\boldsymbol{q}} \tag{3.48}$$

Similarly, choose the Lyapunov equation as shown in equation 错误!未找到引用源。, then $V > 0$, and differentiating V, we can obtain

$$\begin{aligned}\dot{V} &= \frac{1}{2}\dot{\boldsymbol{q}}^{\mathrm{T}}\dot{\boldsymbol{H}}(\boldsymbol{q})\dot{\boldsymbol{q}} + \dot{\boldsymbol{q}}^{\mathrm{T}}\boldsymbol{H}(\boldsymbol{q})\ddot{\boldsymbol{q}} + \Delta\dot{\boldsymbol{q}}^{\mathrm{T}}\boldsymbol{K}\Delta\boldsymbol{q} \\ &= \dot{\boldsymbol{q}}^{\mathrm{T}}[\boldsymbol{I}-\boldsymbol{k}(\boldsymbol{x})]^{-1}\boldsymbol{K}_{\mathrm{Pf}}\Delta\boldsymbol{q} + \dot{\boldsymbol{q}}^{\mathrm{T}}[\boldsymbol{I}-\boldsymbol{k}(\boldsymbol{x})]^{-1}\boldsymbol{K}_{\mathrm{Df}}\Delta\dot{\boldsymbol{q}} - \dot{\boldsymbol{q}}^{\mathrm{T}}\boldsymbol{F}\dot{\boldsymbol{q}} + \Delta\dot{\boldsymbol{q}}^{\mathrm{T}}\boldsymbol{K}\Delta\boldsymbol{q} \\ &= -\Delta\dot{\boldsymbol{q}}^{\mathrm{T}}[\boldsymbol{I}-\boldsymbol{k}(\boldsymbol{x})]^{-1}\boldsymbol{K}_{\mathrm{Pf}}\Delta\boldsymbol{q} - \Delta\dot{\boldsymbol{q}}^{\mathrm{T}}[\boldsymbol{I}-\boldsymbol{k}(\boldsymbol{x})]^{-1}\boldsymbol{K}_{\mathrm{Df}}\Delta\dot{\boldsymbol{q}} - \dot{\boldsymbol{q}}^{\mathrm{T}}\boldsymbol{F}\dot{\boldsymbol{q}} + \Delta\dot{\boldsymbol{q}}^{\mathrm{T}}\boldsymbol{K}\Delta\boldsymbol{q}\end{aligned} \tag{3.49}$$

Choose $\boldsymbol{K} = [\boldsymbol{I} - \boldsymbol{k}(\boldsymbol{x})]^{-1}\boldsymbol{K}_{\mathrm{Pf}}$, then

$$\dot{V} = -\Delta\dot{\boldsymbol{q}}^{\mathrm{T}}[\boldsymbol{I} - \boldsymbol{k}(\boldsymbol{x})]^{-1}\boldsymbol{K}_{\mathrm{Df}}\Delta\dot{\boldsymbol{q}} - \dot{\boldsymbol{q}}^{\mathrm{T}}\boldsymbol{F}\dot{\boldsymbol{q}} \tag{3.50}$$

Because $[\boldsymbol{I} - \boldsymbol{k}(\boldsymbol{x})]^{-1}$ and $\boldsymbol{F}$ are positive definite, as long as we choose $\boldsymbol{K}_{\mathrm{Df}}$ as the positive definite, $\dot{V}$ is negative definite and so the system is asymptotically stable.

3.3.2 Model Identification Using Neural Networks

NNSAC's key part is using neural network to replace the system's inverse model in SAC controller, which involves the neural network model identification, and the BP

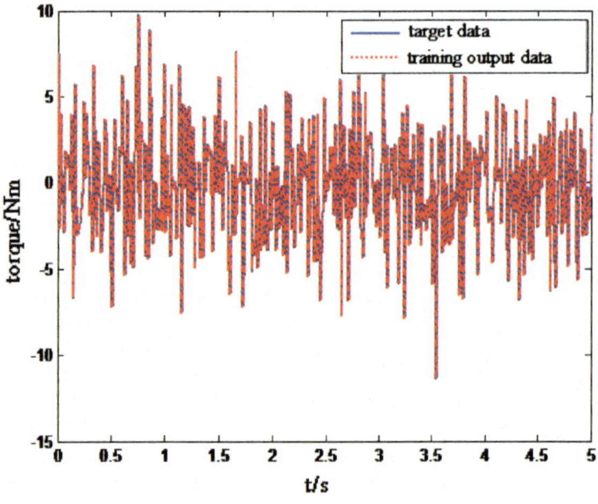

Fig. 3.15 Training data and neural network output

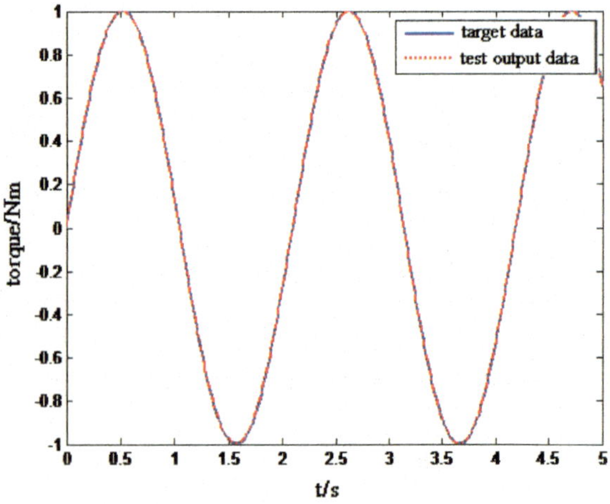

Fig. 3.16 Test data and neural network output

network and RBF network are used more [7]. Because the BP network is a kind of forwardfeedback network, it can achieve arbitrary nonlinear mapping from input to output, and the learning algorithm of BP network belongs to global approximation method, with good generalization ability. Therefore, BP network is used for neural network model identification in this book.

Network's input is the motion angle, angular velocity, and angular acceleration of the exoskeleton suit, and the output is the torque that should be exerted on each joint of the exoskeleton suit. We get the input and output data by using the positive model of the exoskeleton suit; namely, exerting certain torque on each joint of the exoskeleton suit, thus recording the motion information of the exoskeleton suit, then send this information as the input signal into the neural network for training, and the training target is the torque information that was input just now. Because in the actual movement, the force/torque exerted on the exoskeleton suit consists of the force/torque exerted by human, however, this force/torque cannot be measured (or is not measured), so we cannot obtain the network by using the online learning method. Therefore, the off-line learning method is adopted to get network.

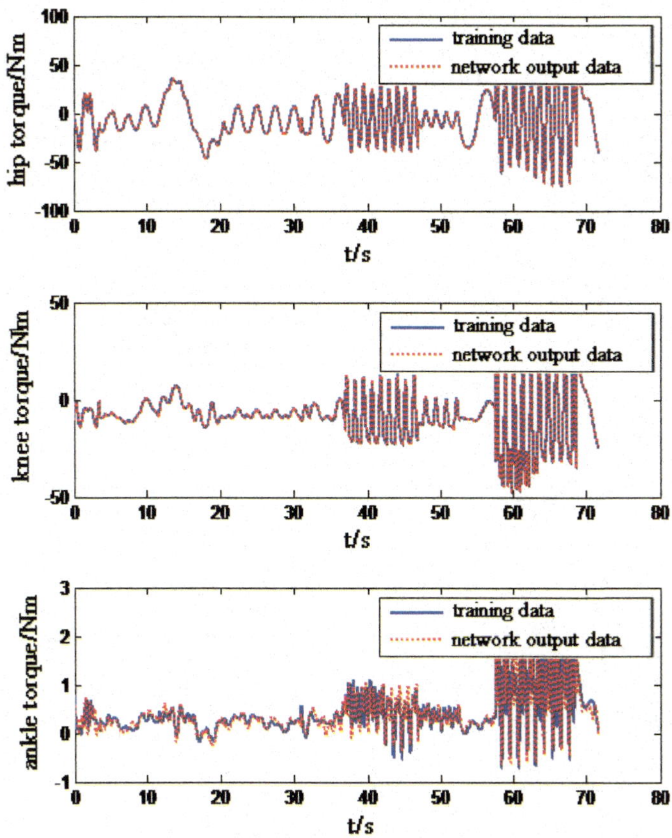

Fig. 3.17 Training data and the neural network's output data

First of all, take the single degree of freedom as the example, and the schematic diagram of the exoskeleton suit of single degree of freedom is as shown in Fig. 3.1. In order to get the training data, first exert a white noise torque signal on the joint of the exoskeleton suit, under the function of which the exoskeleton suit outputs certain movement information (angle, angular velocity, and angular acceleration). Then, use the motion data as the training input data, and the dimensionality is 3; use the torque date as the target data, and the dimensionality is 1; the number of hidden layer neurons of the BP network is set as 6; take the transfer function of input layer and hidden layer as the hyperbolic function and take that of output layer as linear function; and adopt the Levenberg–Marquardt back propagation algorithm as the training method. The training results are as shown in Figs. 3.15 and 3.18. Figure 3.15 shows the training result, and Fig. 3.16 shows the experimental result of test data. The two figures illustrate the effect of neural network model identification is very high, and the network generalization ability is very strong.

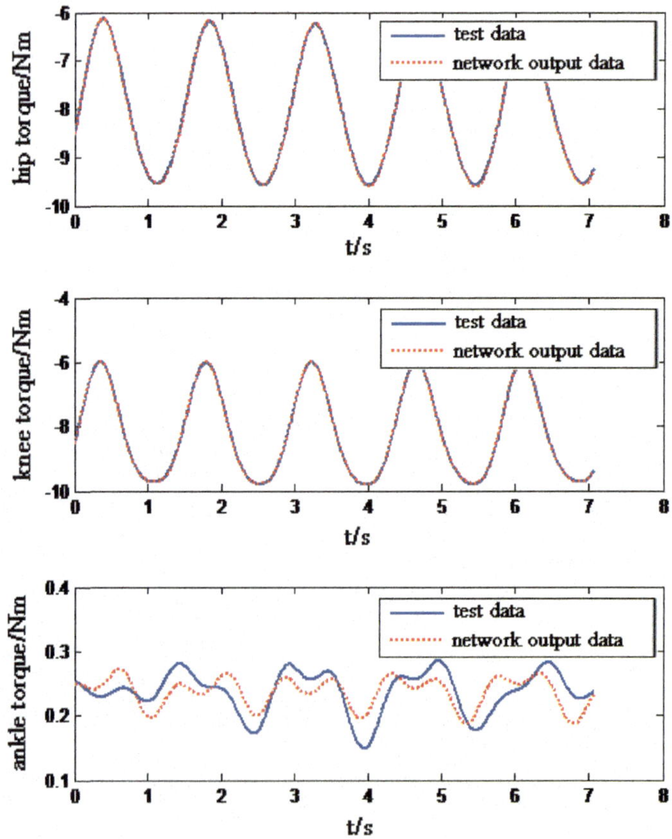

Fig. 3.18 Test data and neural network output

The neural network adopted when it is in single degree of freedom status approximates the system model accurately, and let us look at whether the neural network can accurately approximate the system model when it is in three degrees of freedom status. The model of three degrees of freedom represents the three joints' movement status of lower limb exoskeleton suit, considers three joints' movement information as the neural network input, and considers the three joints' torque as the network output; the method of getting the training data is different from when it is in single degree of freedom status, because in the model of the three degrees of freedom, there is serious coupling among the three joints, so when exerting the torque signal of white noise to each joint, the system instability is easily caused, and a lot of joint movement information is beyond the normal working range, the data being not available, so the inverse model is used to get the training data. Adopting

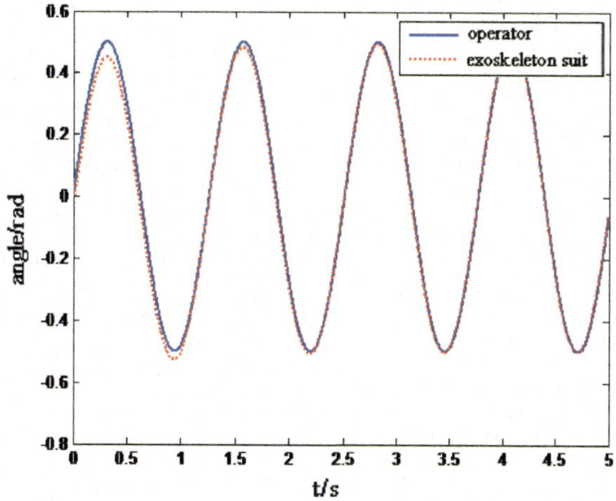

Fig. 3.19 Curve of exoskeleton suit tracking the human body without adding NNSAC control

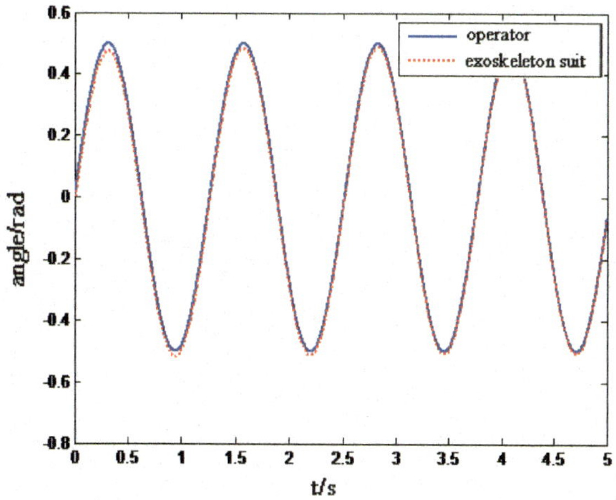

Fig. 3.20 Curve of exoskeleton suit tracking the human body after adding the NNSAC control

the SimMechanics toolbox of MATLAB to establish the mechanical model of the exoskeleton suit, in the process, the system's different trajectories are given. Namely, let the three joints' motion with sinusoidal track, but the three joints's amplitude and frequency of movement are generated randomly each time, and consider the joint cycle of the minimum frequency as simulation time of this time, simulating continuously for 12 times in a row, among which the first 10 groups are used as the training data and the last two groups are used as test data. In the process of generating data, the generated joint movement information should not be beyond the normal motion range of the human body, which should be paid attention to. For example, the frequency should not be too small; otherwise, the simulation time will be overtime, interrupting the simulation. The movement angle of the knee joint should not be less than zero, so consider the 45° as the knee joint's zero point. Also, pay attention to that the angle of the feet and legs should be kept at 90°; namely,

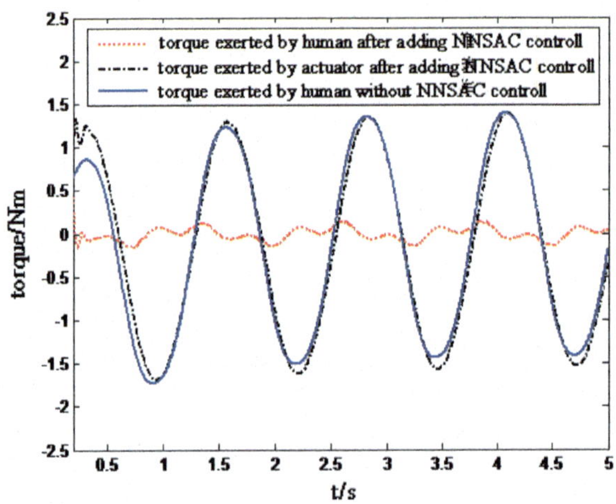

Fig. 3.21 Torque's contrast curve

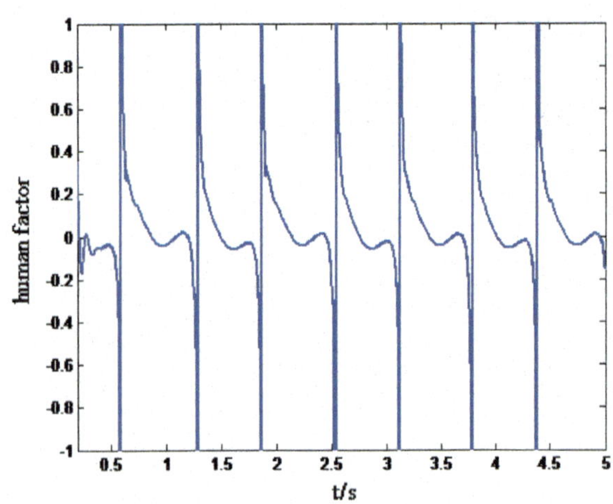

Fig. 3.22 Human factor

pay attention to the transformation relation between the ankle angle defined in the exoskeleton suit model and the ankle angle defined in the SimMechanics. For the network, the BP network is still adopted, whose input is nine dimensions and output is three dimensions; the number of the hidden layer neurons is set as 22; the transfer function and learning algorithm among the layers are same as when it is in single degree of freedom status.

The simulation results are as shown in Figs. 3.17 and 3.18. Figure 3.17 shows the contrast between the target torque data of training data and the torque data of network output. Figure 3.18 shows the contrast between the target torque data of test data and the torque data of network output. From the two figures, we can see that the neural network's approximating level to the system is very high, and the test data also show the network's generalization ability is strong. The trained network can generate the network's Simulink model with the Gensim commands of MATLAB, which being together with the SimMechanics model in Simulink makes simulation based on the sensitivity amplification control method of neural network.

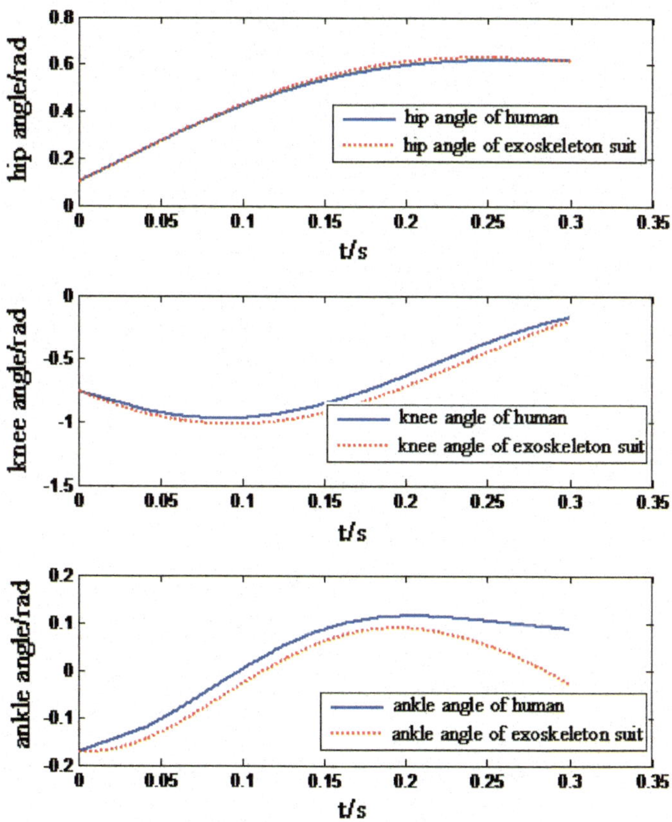

Fig. 3.23 Joint angle tracking curve

3.3.3 Simulation Analysis

3.3.3.1 The Simulation of Single Degree of Freedom

From the previous section, we have got the neural network that can approximate the inverse system model much accurately. According to the principle based on the SAC method of neural network shown in Fig. 3.14, the control loop that has introduced the neural network into this system can control the movement of the exoskeleton suit. First of all, take the system of single degree of freedom as an example for verification. When simulating, take $\alpha = 10$; the operator's desired trajectory is sine curve whose amplitude is 0.5 rad and frequency is 5 Hz. Figure 3.19 shows the angle curve of exoskeleton suit tracking the human body without adding NNSAC control, and Fig. 3.20 shows the angle curve of exoskeleton suit tracking the human body when the NNSAC control has been

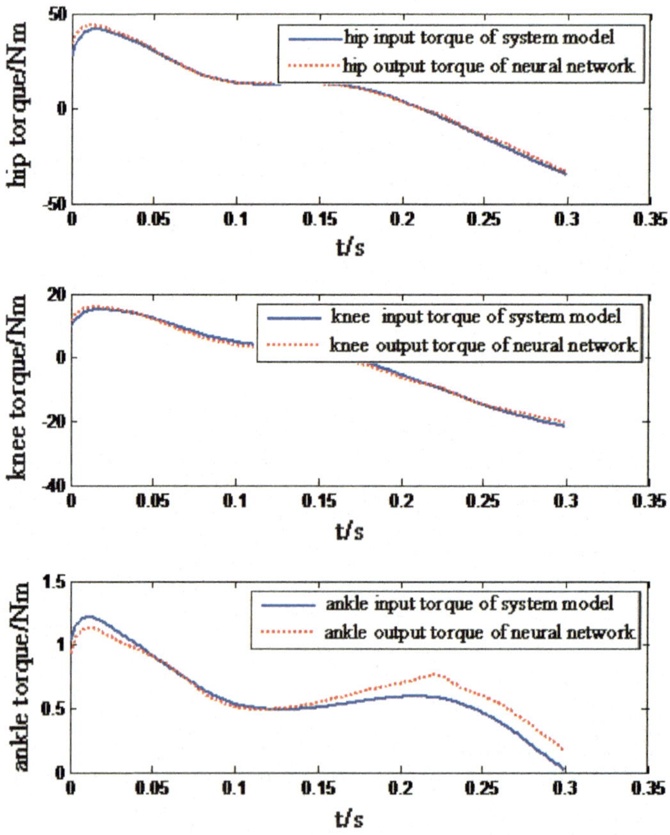

Fig. 3.24 System's input torque and the neural network's output torque

added. The two figures illustrate that the lower limb exoskeleton suit can track the movement trajectory of the body in both cases, and the tracking error is very small.

Figure 3.21 shows the contrast curve between the torque of human and the machine in both cases. The solid line in the figure represents the torque exerted by the operator without adding NNSAC control, when all the torque required by the exoskeleton suit movement is exerted by the operator; as a result, the operator's energy consumption is very large. The dotted line in the figure represents the torque exerted by the operator after the NNSAC control has been added, and from the figure, we can see that the torque exerted by the operator becomes very small, when most of the torque required by the exoskeleton suit movement is provided by the NNSAC controller (actuator); as shown by the chain dotted line in the figure, the energy consumption of the human body has been greatly reduced. Because the torque of actuator output is zero when without NNSAC control, it is not shown in figure. Figure 3.22 shows the human factor *HF* of the system, and we can see that most of the *HF* is restricted within ±0.3. Comparing Fig. 3.21 with Fig. 3.22, we can see that the *HF*'s relatively large peak points in Fig. 3.22 correspond to the

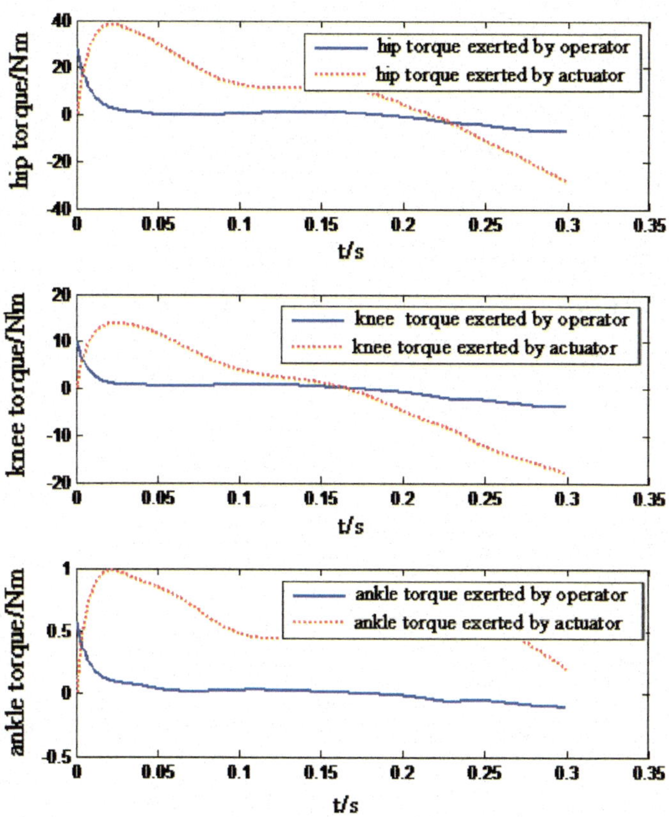

Fig. 3.25 Torque exerted by the operator and the torque exerted by the actuator

ZCP of resultant external force in Fig. 3.21, when *HF* is close to or even more than 1, but we can see that the acting force exerted by human is very small at this time through observing Fig. 3.21. The system's maximum human factor *HFM* is 0.3099.

3.3.3.2 The Simulation of Swing Leg

The condition of the simulation of swing leg is consistent with the sensitivity amplification simulation, and the Sect. 3.2.5.2 can be referred to. The simulation results are as shown in Figs. 3.23, 3.24, 3.25 and 3.26. Figure 3.23 shows the joint angle tracking curve, illustrating that the exoskeleton suit can track the movement trajectory of the human body well. Figure 3.24 shows the contrast curve between the torque required by the movement of the system and the neural network's output torque, illustrating that the system's inverse model obtained by neural network can track the system's input torque well. Figure 3.25 shows the contrast curve between

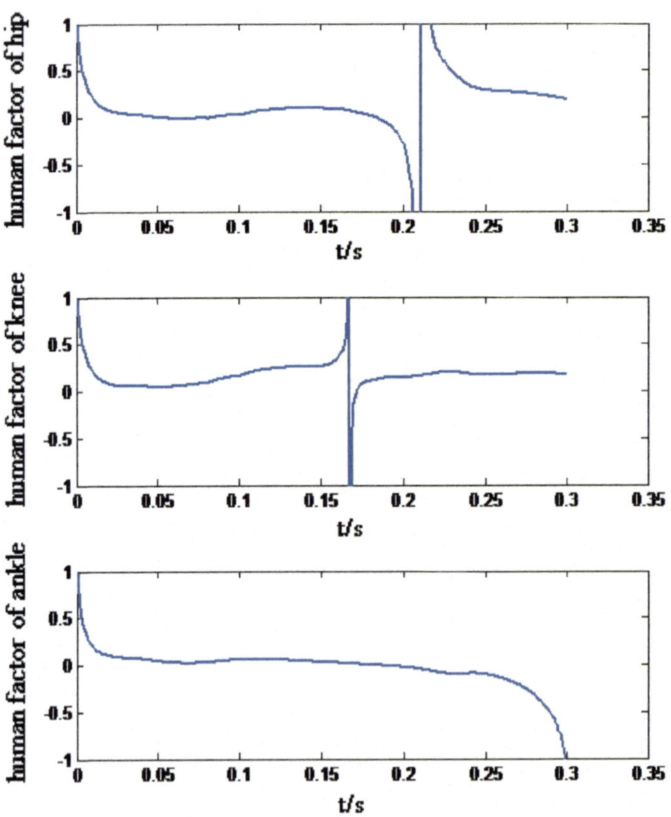

Fig. 3.26 Human factor of joint space

the human-machine acting torque on each joint exerted by the operator and the control torque exerted by the actuator, illustrating that in the movement process, human only needs to provide certain starting torque; after the movement has been started, most of the torque is provide by the actuator, and the torque provided by human is only limited to changing the movement state, illustrating that the NNSAC has reduced the human-machine acting force, thereby reducing the energy consumption of the human body, reaching the predetermined result.

Figure 3.26 shows the human factor of joint space, and we can see that most of the human factor $\boldsymbol{HF}$ of the three joints is limited within ± 0.3. The reason why there are only two peaks in each joint curve is because there are only two ZCPs of the resultant external force. The three joints's maximum human factor is $\boldsymbol{HFM} = [0.1674 \quad 0.1748 \quad 0.1085]$. The above description illustrates that the designed controller is effective and feasible.

References

1. Kazerooni, H, Steger R. The Berkeley Lower Extremity Exoskeleton [J]. Journal of Dynamic Systems, Measurement, and Control, 2005, 128 (3): 14–25.
2. Li Xiaoming. The Robot's Remote Control Based on the Exoskeleton Technology [D]. Zhejiang University, Doctoral Dissertation, Hangzhou: 2004.
3. Huang, Lihua. Robotics Locomation Control [D]. University of California, Ph. D, Berkeley: 2005.
4. Ghan, Justim, Ryan Steger, Kazerooni H. Control and System Identification for the Berkeley Lower Extremity Exoskeleton(BLEEX) [J]. Advanced Robotics, 2006, 20 (9): 989–1014.
5. Kazerooni, H, Huang L H, Steger R. On the Control of the Berkeley lower extremity exoskeleton (BLEEX) [C]. Proceedings of IEEE International Conference on Robotics and Automation (ICRA 2005), Barcelona, Spain, 2005: 4353–4360.
6. Winter, D A. Biomechanics of Human Movement [M]. New York: John Wiley and Sons, 1979.
7. Sun Zengqi, Zhang Zaixing, Deng Zhidong. Intelligent Control Theory and Techniques [M]. Beijing: Tsinghua University Press, 1997.

Chapter 4
Direct Force Control of Exoskeleton Suit

4.1 Introduction

Except relying heavily on the precise dynamic model of the system, sensitivity amplification control method's another big defect is that the exoskeleton suit itself cannot judge how the control effect is. This is because the information that the exoskeleton suit can obtain only is the movement information itself, and how much force is exerted on the exoskeleton suit by human, namely, the magnitude of the energy consumed by the human body can not be obtained by the exoskeleton suit. When debugging the exoskeleton suit, only according to the feeling of people, adjust the controller's parameter of exoskeleton suit constantly, which makes the exoskeleton suit not be able to adjust the control torque according to the magnitude of the force exerted by human, let alone training; when the load changes, the controller parameter need be adjusted, which is inconvenient. Therefore, using the force sensor to measure the acting force on the exoskeleton suit exerted by human, making the controller to be able to control the exoskeleton suit's movement according to the magnitude of the force, and making the force to be maintained in lower level are the most direct idea. And force control technology of using the force sensor to measure the acting force between the machine and environment and to control them has been extensively adopted in the control of mechanical arms and robots. Learning from the force control technology of mechanical arms and robots and applying them to the exoskeleton suit's control are a good exploration.

The force control of the robot refers to that in the robot's movement control of the limited space, and the requirement for the restricted movement control of the acting force of the robot's extremity contacting with the outside world has been added. The force control of the robot is generally applied to the situation in which the robot contacts with the environment, controlling the contact force between the robot and environment. For example, the robot is used for assembly, cutting, polishing, grinding, scrubbing, deburring, etc. [1–5]. The study of the robot's force control has already lasted nearly half a century; the control methods that have been

Z. Yang et al., *Force Control Theory and Method of Human Load Carrying Exoskeleton Suit*, DOI 10.1007/978-3-662-54144-9_4

put forward are also varied [5–8], and the literatures [3, 9, 10] have described force control methods of the robot in detail, so we will not give unnecessary details in this book. After analyzing the above documents comprehensively, we can get the basic structure of the robot's force control as shown in Fig. 4.1.

In Fig. 4.1, $\boldsymbol{x}, \dot{\boldsymbol{x}}$ represents the position and speed of the robot's extremity in the operation space, and $\boldsymbol{x}_{\mathrm{i}}$ represents the position of the environment. When the environment and the robot's extremity generate the position error (generally one way), the environment will produce contact force $\boldsymbol{f}$ acting on the robot. Among them, $\boldsymbol{K}_{\mathrm{E}}$ represents the environment model, $\boldsymbol{x}_{\mathrm{d}}, \dot{\boldsymbol{x}}_{\mathrm{d}}, \boldsymbol{f}_{\mathrm{d}}$ represent the desired robot's terminal position, speed, and environmental acting force, respectively.

The exoskeleton suit's force control structure can be summarized as shown in Fig. 4.2. In the figure, $\boldsymbol{x}_{\mathrm{h}}$ represents the human movement trajectory in the operation space, $\boldsymbol{G}_{\mathrm{h}}$ represents the human-machine interaction model; the desired position of exoskeleton suit's extremity is $\boldsymbol{x}_{\mathrm{d}} = \boldsymbol{x}_{\mathrm{h}}$, and the desired speed is

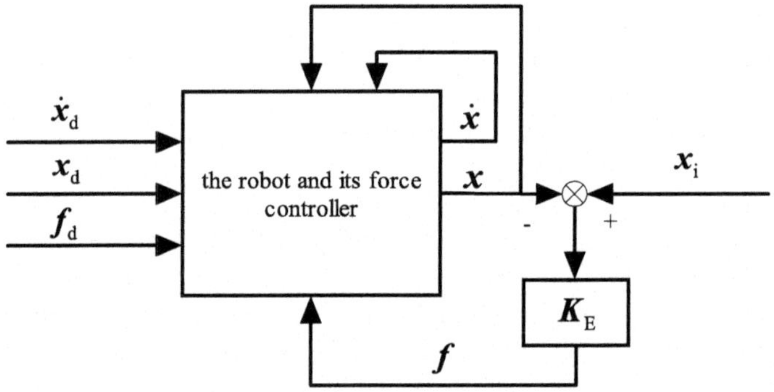

Fig. 4.1 Basic structure of the robot's force control

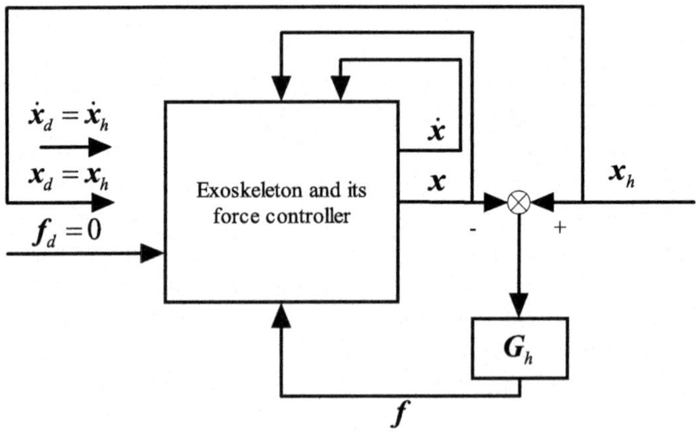

Fig. 4.2 Basic structure of the exoskeleton suit's force control

$\dot{x}_d = \dot{x}_h$, but the controller cannot be obtained; therefore, it is not shown in the figure.

By contrasting Fig. 4.1 with Fig. 4.2, we can see that there is something in common and something different between the force control of the exoskeleton suit and that of the robot. Their common points are as follows:

1. They both have force feedback information;
2. They both are accompanied with movement feedback information, and generally, the feedback information of joint space can be converted into movement information of operation space when needed.

Their main differences are as follows:

1. Generally, in the robot's force control, the desired position $\boldsymbol{x}_d$ and speed $\dot{\boldsymbol{x}}_d$ need to be known and determined in advance, while in the force control of the exoskeleton suit, the controller cannot directly obtain desired position $\boldsymbol{x}_d$ and the rate signal $\dot{\boldsymbol{x}}_d$ (even if there are methods to obtain, they are not suitable for being used in the exoskeleton suit; for example, installing the sensor on human body to measure its movement position and velocity information, which are used as the desired position and rate signal of the exoskeleton suit).
2. Generally, there is also the desired contact force signal $\boldsymbol{f}_d$ in the robot's force control, and in the force control of the exoskeleton suit, the desired contact force is $\boldsymbol{f}_d = 0$.
3. In the robot's force control, the contact force $\boldsymbol{f}$ is from the position error $\boldsymbol{x}_i - \boldsymbol{x}$ between the environment and the robot's end position, and the position of the environment has nothing to do with the robot's desired position $\boldsymbol{x}_d$. While in the force control of the exoskeleton suit, human is the environment; the contact force is generated from the position error $\boldsymbol{x}_h - \boldsymbol{x}$ between human and the end position of the exoskeleton suit; human's real time position is the desired position $\boldsymbol{x}_d = \boldsymbol{x}_h$ of the exoskeleton suit; and the velocity and acceleration are the same.
4. In the robot's force control, the environmental position $\boldsymbol{x}_i$ is fixed in general, and the so-called environmental uncertainty refers to the information of environmental surface shape, soft degree, etc. that cannot be obtained in advance. While in the force control of the exoskeleton suit, human is the environment, and human's position $\boldsymbol{x}_h$ is always changing.
5. In the robot's force control, generally, the motion control is in free space (the robot's extremity is far away from the environment), and the force control is in the constraint space (the robot's extremity contacts with the environment). While the force control of the exoskeleton suit is always in the constraint space, the exoskeleton suit contacts with the environment from beginning to end.

These differences make the force control of the robot not to be able to directly transplant into the exoskeleton suit system, and it must be redesigned according to the characteristics of the exoskeleton suit. The several force control methods of exoskeleton suit introduced in this book are to redesign the force controller according to the characteristics of the exoskeleton suit, whose characteristic is the smaller human-machine contact force (near zero) that controls the movement of the exoskeleton suit to comply with human, called zero force compliance control.

This chapter will introduce the direct force control of the exoskeleton suit. In the interaction control strategy of the robot/mechanical arms and the environment, the direct force control refers to the control of reaching the desired value by using the force feedback loop to control the contact force directly, as opposed to indirect force control. Their main difference is that the indirect force control obtains the force control through the motion control, with explicit force feedback loop, and the Chaps. 5–8 belong to the category of indirect force control.

Direct force control adjusts the force directly similar to the explicit force control based on the force in the so-called explicit force control method [3, 5]. The direct force control method of the exoskeleton suit introduced in this chapter uses the information of the force/torque exerted by human on the exoskeleton suit in the multidimensional force/torque sensor operation space, and according to the information and the static model of the exoskeleton suit, construct the control torque of the joint space, control the joint movement of the exoskeleton suit directly, and then realize the minimization of the contact force between the exoskeleton suit's extremity and human in the operation space.

4.2 Direct Force Control of Exoskeleton Suit

Referring to Eq. (2.74), we can get the dynamic model of the exoskeleton suit system

$$\boldsymbol{H}(\boldsymbol{q})\ddot{\boldsymbol{q}} + \boldsymbol{C}(\boldsymbol{q},\dot{\boldsymbol{q}})\dot{\boldsymbol{q}} + \boldsymbol{F}\dot{\boldsymbol{q}} + \boldsymbol{G}(\boldsymbol{q}) = \boldsymbol{T}_{\mathrm{a}} + \boldsymbol{T}_{\mathrm{hm}} \tag{4.1}$$

In the equation, $\boldsymbol{H}(\boldsymbol{q})$ is the inertia matrix; $\boldsymbol{C}(\boldsymbol{q},\dot{\boldsymbol{q}})$ is the Coriolis term; $\boldsymbol{F}$ represents the movement friction coefficient; $\boldsymbol{G}(\boldsymbol{q})$ is the gravity term; $\boldsymbol{T}_{\mathrm{a}}$ represents the torque exerted by the actuator; $\boldsymbol{T}_{\mathrm{hm}}$ represents the human-machine moment, namely the moment exerted by the operator. Without considering other disturbance torques, the sum of $\boldsymbol{T}_{\mathrm{a}}$ and $\boldsymbol{T}_{\mathrm{hm}}$ is the resultant moment acting on the exoskeleton suit and

$$\boldsymbol{T}_{\mathrm{hm}} = \boldsymbol{J}^{\mathrm{T}}(\boldsymbol{q})\boldsymbol{f} \tag{4.2}$$

In the equation, $\boldsymbol{J}$ represents the Jacobian matrix of the exoskeleton suit; $\boldsymbol{f}$ represents the generalized force of human acting on the trunk of the exoskeleton suit in the operation space; and for the definition of $\boldsymbol{J}$ and $\boldsymbol{f}$, please refer to the Sect. 2.3.2.

The principle diagram of the direct force control of the exoskeleton suit is shown in Fig. 4.3. Among them, G_{a} represents the exoskeleton suit's kinetic model, namely Eq. (4.1), and G_{h} represents the human-machine interaction model. Design the control law as

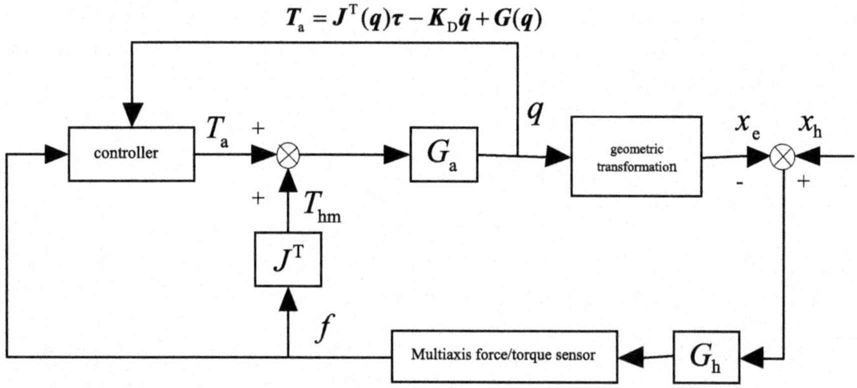

Fig. 4.3 Principle diagram of the direct force control of the exoskeleton suit

$$T_a = J^T(q)\tau - K_D\dot{q} + G(q) \tag{4.3}$$

In the equation, τ represents the generalized control force that can be designed and exerted by the actuator, namely the control force and control torque in the operation space; $G(q)$ is used to compensate the static gravity moment; K_D is a positive definite gain matrix, representing the damping coefficient matrix; and $-K_D\dot{q}$ provides an additional damping torque to improve the dynamic response of the system.

Substituting Eqs. (4.3) and (4.2) into Eq. (4.1), and we can get

$$H(q)\ddot{q} + C(q,\dot{q})\dot{q} + (F + K_D)\dot{q} = J^T(q)\tau + J^T(q)f \tag{4.4}$$

Based on the proportional control, design τ as

$$\tau = K_f f \tag{4.5}$$

In the equation, f can be obtained by measuring the multidimensional force/torque sensor; K_f represents the gain of the controller.

Then, according to Eq. (4.5), Eq. (4.4) can be changed to

$$\tau = K_f f \tag{4.6}$$

4.3 Stability Analysis

In order to analyze the system's stability expressed by Eq. (4.6), we need to know the mathematical expression of the human-machine interaction model. As mentioned in Sect. 2.5, the linear model of human-machine interaction can be expressed by the mathematical model shown in equations from (2.106) to (2.108). If we define $\Delta \boldsymbol{x}_{\rm he} = \boldsymbol{x}_{\rm h} - \boldsymbol{x}_{\rm e}$, $\Delta \dot{\boldsymbol{x}}_{\rm he} = \dot{\boldsymbol{x}}_{\rm h} - \dot{\boldsymbol{x}}_{\rm e}$, $\Delta \ddot{\boldsymbol{x}}_{\rm he} = \ddot{\boldsymbol{x}}_{\rm h} - \ddot{\boldsymbol{x}}_{\rm e}$ to represent the motion error between human and the exoskeleton suit, respectively, the equations from (2.106) to (2.108) can be reexpressed as

$$\boldsymbol{f} = \boldsymbol{K}_{\rm Pf}\Delta \boldsymbol{x}_{\rm he} + \boldsymbol{K}_{\rm Df}\Delta \dot{\boldsymbol{x}}_{\rm he} + \boldsymbol{K}_{\rm Mf}\Delta \ddot{\boldsymbol{x}}_{\rm he} \tag{4.7}$$

$$\boldsymbol{f} = \boldsymbol{K}_{\rm Pf}\Delta \boldsymbol{x}_{\rm he} + \boldsymbol{K}_{\rm Df}\Delta \dot{\boldsymbol{x}}_{\rm he} \tag{4.8}$$

$$\boldsymbol{f} = \boldsymbol{K}_{\rm Pf}\Delta \boldsymbol{x}_{\rm he} \tag{4.9}$$

The model expressed by Eq. (4.7) is called impedance model. The magnitude of the spring coefficient matrix, damping coefficient matrix, and inertia coefficient matrix not only is different for different individual operators, but also is different for the same operator at different time. But when making the stability analysis, we assume that they are constant. In the following part, we will use three human-machine interaction models, respectively, to analyze the stability of the designed controller.

4.3.1 Spring Model

After substituting Eq. (4.9) into Eq. (4.6), we can get the system dynamics

$$\boldsymbol{H}(\boldsymbol{q})\ddot{\boldsymbol{q}} + \boldsymbol{C}(\boldsymbol{q},\dot{\boldsymbol{q}})\dot{\boldsymbol{q}} + (\boldsymbol{F} + \boldsymbol{K}_{\rm D})\dot{\boldsymbol{q}} = \boldsymbol{J}^{\rm T}(\boldsymbol{q})(\boldsymbol{I} + \boldsymbol{K}_{\rm f})\boldsymbol{K}_{\rm Pf}\Delta \boldsymbol{x}_{\rm he} \tag{4.10}$$

Let

$$\boldsymbol{K}_1 = (\boldsymbol{I} + \boldsymbol{K}_{\rm f})\boldsymbol{K}_{\rm Pf} \tag{4.11}$$

Then, the stability of the system can be proved by introducing the Lyapunov equation shown in Eq. (4.12)

$$V = \frac{1}{2}\dot{\boldsymbol{q}}^{\rm T}\boldsymbol{H}(\boldsymbol{q})\dot{\boldsymbol{q}} + \frac{1}{2}\Delta \boldsymbol{x}_{\rm he}{}^{\rm T}\boldsymbol{K}_1\Delta \boldsymbol{x}_{\rm he} \tag{4.12}$$

Through Eq. (4.12), we can see that $V > 0$; by differentiating V, then we can get

$$\begin{aligned}\dot{V} &= \frac{1}{2}\dot{\boldsymbol{q}}^{\rm T}\dot{\boldsymbol{H}}(\boldsymbol{q},\dot{\boldsymbol{q}})\dot{\boldsymbol{q}} + \dot{\boldsymbol{q}}^{\rm T}\boldsymbol{H}(\boldsymbol{q})\ddot{\boldsymbol{q}} + \Delta \dot{\boldsymbol{x}}_{\rm he}{}^{\rm T}\boldsymbol{K}_1\Delta \boldsymbol{x}_{\rm he} \\ &= -\dot{\boldsymbol{q}}^{\rm T}(\boldsymbol{F} + \boldsymbol{K}_{\rm D})\dot{\boldsymbol{q}}\end{aligned} \tag{4.13}$$

As both matrix $\boldsymbol{F}$ and $\boldsymbol{K}_{\rm D}$ are positive definite, $\dot{V}$ is not positive, and so the system is asymptotically stable in the balance point.

4.3.2 Spring Damping Model

After substituting Eq. (4.8) into Eq. (4.6), we can get the system dynamics

$$\boldsymbol{H}(\boldsymbol{q})\ddot{\boldsymbol{q}} + \boldsymbol{C}(\boldsymbol{q}, \dot{\boldsymbol{q}})\dot{\boldsymbol{q}} + (\boldsymbol{F} + \boldsymbol{K}_{\mathrm{D}})\dot{\boldsymbol{q}} = \boldsymbol{J}^{\mathrm{T}}(\boldsymbol{q})(\boldsymbol{K}_1 \Delta \boldsymbol{x}_{\mathrm{he}} + \boldsymbol{K}_2 \Delta \dot{\boldsymbol{x}}_{\mathrm{he}}) \quad (4.14)$$

In the equation

$$\boldsymbol{K}_1 = (\boldsymbol{I} + \boldsymbol{K}_{\mathrm{f}})\boldsymbol{K}_{\mathrm{Pf}} \quad (4.15)$$

$$\boldsymbol{K}_2 = (\boldsymbol{I} + \boldsymbol{K}_{\mathrm{f}})\boldsymbol{K}_{\mathrm{Df}} \quad (4.16)$$

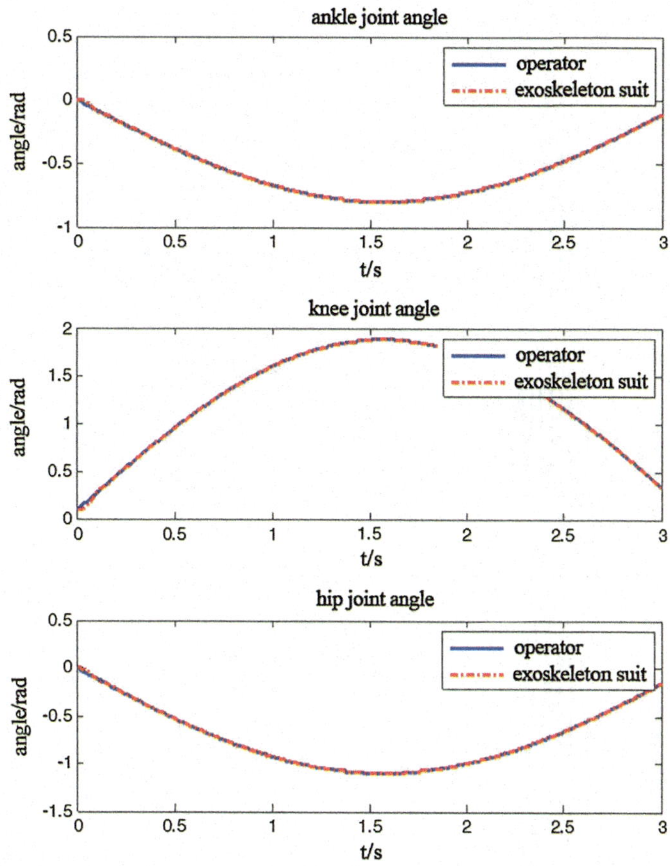

Fig. 4.4 Trajectory tracking curve

Similarly, by introducing the Lyapunov function, we have

$$V = \frac{1}{2}\dot{\boldsymbol{q}}^{\mathrm{T}}\boldsymbol{H}(\boldsymbol{q})\dot{\boldsymbol{q}} + \frac{1}{2}\Delta\boldsymbol{x}_{\mathrm{he}}^{\mathrm{T}}\boldsymbol{K}_1\Delta\boldsymbol{x}_{\mathrm{he}} \tag{4.17}$$

Then, we can see that $V > 0$; after differentiating V, we can get

$$\begin{aligned}\dot{V} &= \frac{1}{2}\dot{\boldsymbol{q}}^{\mathrm{T}}\dot{\boldsymbol{H}}(\boldsymbol{q},\dot{\boldsymbol{q}})\dot{\boldsymbol{q}} + \dot{\boldsymbol{q}}^{\mathrm{T}}\boldsymbol{H}(\boldsymbol{q})\ddot{\boldsymbol{q}} + \Delta\dot{\boldsymbol{x}}_{\mathrm{he}}^{\mathrm{T}}\boldsymbol{K}_1\Delta\boldsymbol{x}_{\mathrm{he}} \\ &= -\Delta\dot{\boldsymbol{x}}_{\mathrm{he}}^{\mathrm{T}}\boldsymbol{K}_2\Delta\dot{\boldsymbol{x}}_{\mathrm{he}} - \dot{\boldsymbol{q}}^{\mathrm{T}}(\boldsymbol{F}+\boldsymbol{K}_{\mathrm{D}})\dot{\boldsymbol{q}}\end{aligned} \tag{4.18}$$

As both matrix $\boldsymbol{F}$ and $\boldsymbol{K}_{\mathrm{D}}$ are positive definite, therefore, as long as the chosen matrix $\boldsymbol{K}_2$ satisfies the positive definiteness, the $\dot{V}$ is not positive, and so the system is asymptotically stable in the balance point.

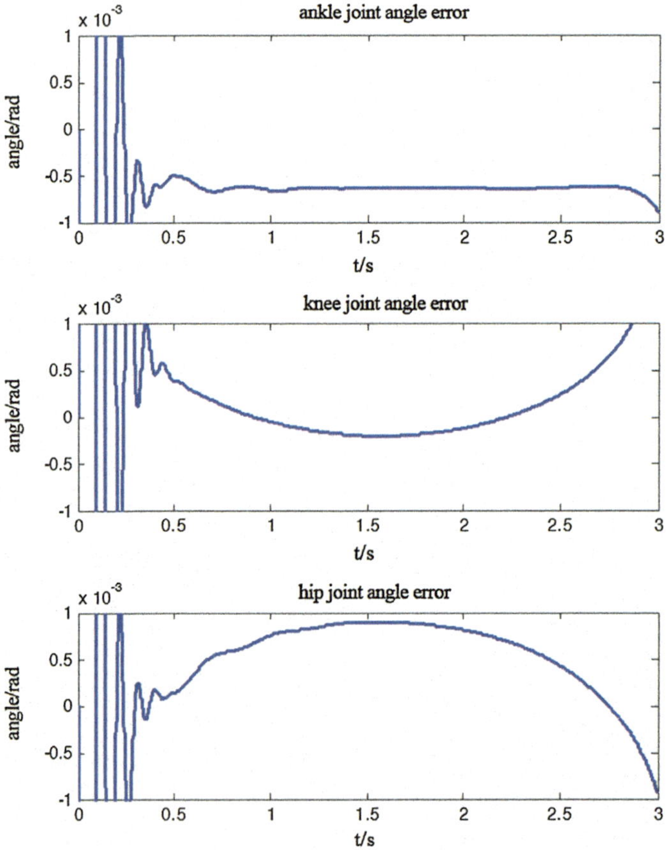

Fig. 4.5 Trajectory tracking error

4.3.3 Impedance Model

After substituting Eq. (4.7) into Eq. (4.6), we can get the system dynamics

$$\boldsymbol{H}(\boldsymbol{q})\ddot{\boldsymbol{q}}+\boldsymbol{C}(\boldsymbol{q},\dot{\boldsymbol{q}})\dot{\boldsymbol{q}}+(\boldsymbol{F}+\boldsymbol{K}_{\mathrm{D}})\dot{\boldsymbol{q}}=\boldsymbol{J}^{\mathrm{T}}(\boldsymbol{q})(\boldsymbol{K}_1\Delta\boldsymbol{x}_{\mathrm{he}}+\boldsymbol{K}_2\Delta\dot{\boldsymbol{x}}_{\mathrm{he}}+\boldsymbol{K}_3\Delta\ddot{\boldsymbol{x}}_{\mathrm{he}}) \tag{4.19}$$

In the equation, $\boldsymbol{K}_1$ and $\boldsymbol{K}_2$ are as shown in Eq. (4.15) and Eq. (4.16); $\boldsymbol{K}_3$ is as shown in Eq. (4.20)

$$\boldsymbol{K}_3=(\boldsymbol{I}+\boldsymbol{K}_{\mathrm{f}})\boldsymbol{K}_{\mathrm{Mf}} \tag{4.20}$$

Similarly, by introducing the Lyapunov function, we have

$$V=\frac{1}{2}\dot{\boldsymbol{q}}^{\mathrm{T}}\boldsymbol{H}(\boldsymbol{q})\dot{\boldsymbol{q}}+\frac{1}{2}\Delta\boldsymbol{x}_{\mathrm{he}}{}^{\mathrm{T}}\boldsymbol{K}_1\Delta\boldsymbol{x}_{\mathrm{he}}+\frac{1}{2}\Delta\dot{\boldsymbol{x}}_{\mathrm{he}}{}^{\mathrm{T}}\boldsymbol{K}_3\Delta\dot{\boldsymbol{x}}_{\mathrm{he}} \tag{4.21}$$

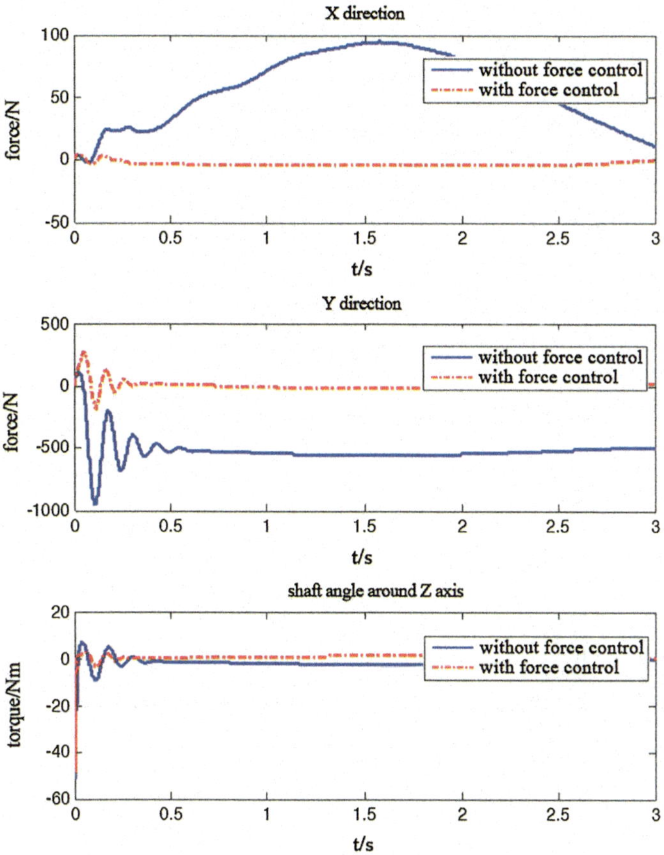

Fig. 4.6 Human-machine interacting force in operation space

Then, we can see that $V > 0$; after differentiating V, we can get

$$\begin{aligned}\dot{V} &= \frac{1}{2}\dot{\boldsymbol{q}}^{\mathrm{T}}\dot{\boldsymbol{H}}(\boldsymbol{q},\dot{\boldsymbol{q}})\dot{\boldsymbol{q}} + \dot{\boldsymbol{q}}^{\mathrm{T}}\boldsymbol{H}(\boldsymbol{q})\ddot{\boldsymbol{q}} + \Delta\dot{\boldsymbol{x}}_{\mathrm{he}}{}^{\mathrm{T}}\boldsymbol{K}_1\Delta\boldsymbol{x}_{\mathrm{he}} + \Delta\dot{\boldsymbol{x}}_{\mathrm{he}}{}^{\mathrm{T}}\boldsymbol{K}_3\Delta\ddot{\boldsymbol{x}}_{\mathrm{he}} \\ &= -\Delta\dot{\boldsymbol{x}}_{\mathrm{he}}{}^{\mathrm{T}}\boldsymbol{K}_2\Delta\dot{\boldsymbol{x}}_{\mathrm{he}} - \dot{\boldsymbol{q}}^{\mathrm{T}}(\boldsymbol{F} + \boldsymbol{K}_{\mathrm{D}})\dot{\boldsymbol{q}}\end{aligned} \tag{4.22}$$

As both matrix $\boldsymbol{F}$ and $\boldsymbol{K}_{\mathrm{D}}$ are positive definite, therefore, as long as the chosen matrix $\boldsymbol{K}_2$ satisfies the positive definiteness, the $\dot{V}$ is not positive, and so the system is asymptotically stable in the balance point.

After contrasting, we can see that Eq. (4.8) is special form of Eq. (4.7), and Eq. (4.9) is special form of Eq. (4.8); so as long as the stability of impedance model is proved, the spring model and spring damping model are also stable, and the above evidence also verifies this point.

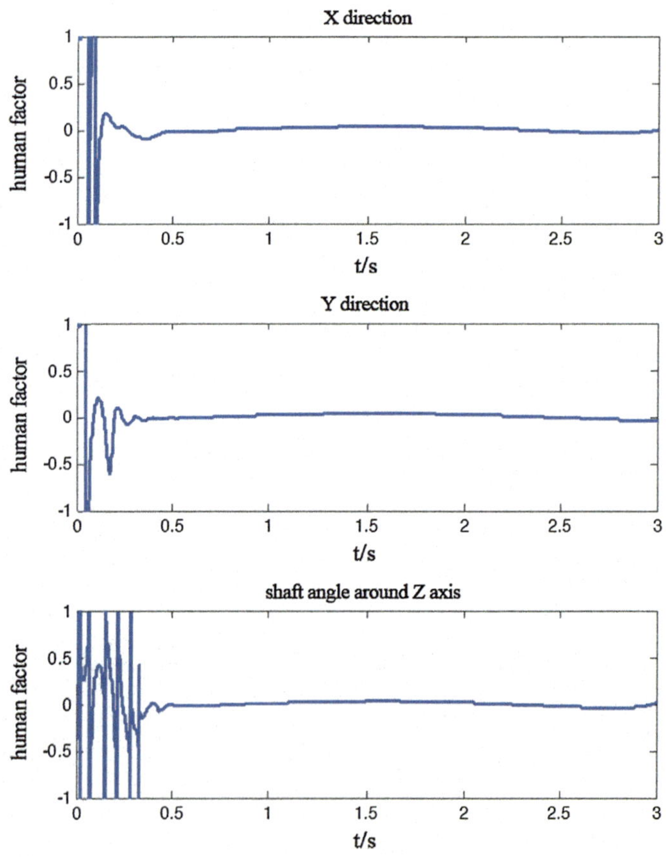

Fig. 4.7 Human factor in operation space

4.4 Simulation

In order to verify the effectiveness of the designed control law, take the proportional controller shown in Eq. (4.5) and the spring damping human-machine interaction model shown in Eq. (4.8) as examples to present the actual simulation examples. Taking the support leg model as the simulation object shown in Fig. 2.5a, use the SimMechanics toolbox of MATLAB to establish the mechanical model of the exoskeleton suit and human, and they both are linked together in the part of the trunk's center of gravity. The mass and geometric attributes of human and exoskeleton suit adopt the human body parameter of Winter [11]. They both are linked together in the part of the trunk's center of gravity through a Custom Joint with Joint Spring & Damper (JSD). So JSD establishes the interaction relationship (coupling relationship) between the exoskeleton suit and the operator, and JSD also can set their spring coefficient matrixes and damping coefficient matrixes in

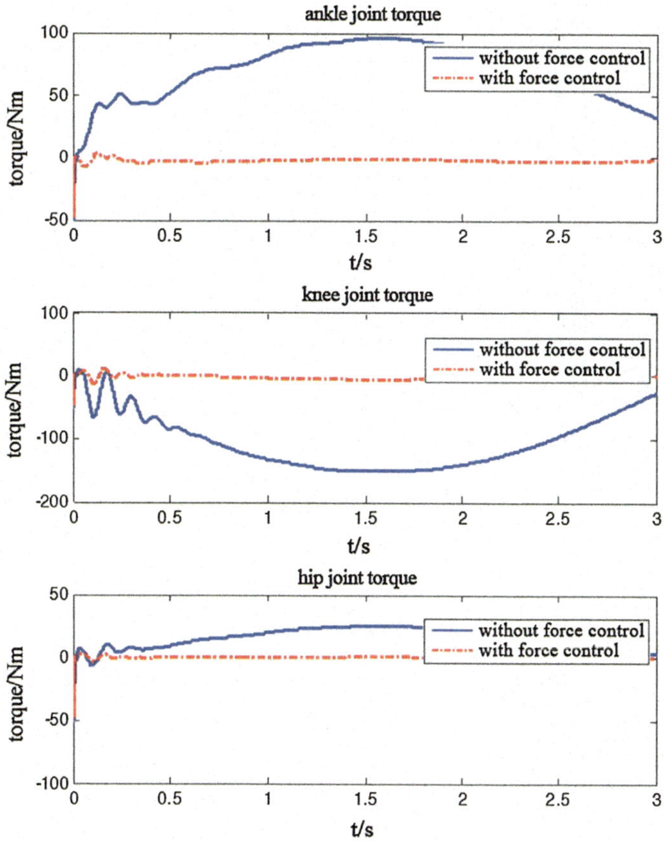

Fig. 4.8 Human-machine interacting torque of joint space

three-dimensional space. The sensor can be added to the universal joint to measure the interaction force f between them, the equivalent of multidimensional force/torque sensor's function. In this simulation, the spring damping model shown

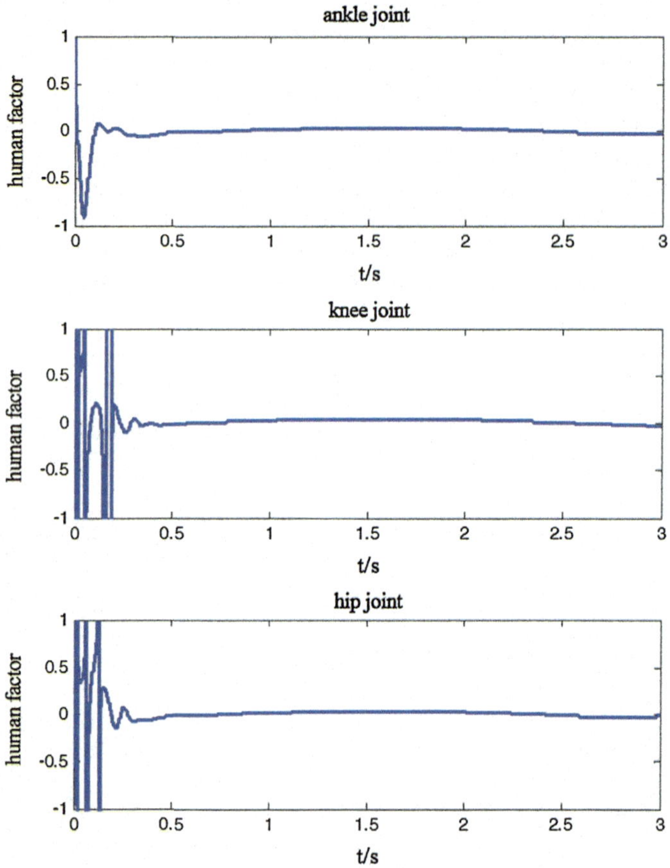

Fig. 4.9 Human factor in joint space

Table 4.1 Human-machine interaction model parameters

Parameters	Parameter 1	Parameter 2	Parameter 3	Parameter 4
K_{pfx}	10,000	5000	2500	1250
K_{dfx}	200	100	50	25
K_{pfy}	15,000	10,000	5000	2500
K_{dfy}	1000	500	250	125
K_{pfrz}	500	250	125	62.5
K_{dfrz}	10	5	2.5	1.25

in Eq. (4.8) is used to describe the relationship between $\boldsymbol{f}$ and the human-machine error, among which the spring coefficient $\boldsymbol{K}_{Pf}$ and damping coefficient $\boldsymbol{K}_{Df}$ are set in JSD.

Take the action of the human body driving the exoskeleton suit to squat as the example to simulate. Because there is no real human squatting movement data, so first of all, through the method of simulation experiment to get the data of body squatting movement. After continuous simulation trial, obtain the approximation of human body squatting movement data. The specifics are as follows:

$$q_{h1} = \frac{4}{5}\sin(t+\pi) - \frac{5}{180}\pi \tag{4.23}$$

$$q_{h2} = \frac{9}{5}\sin(t) + \frac{5}{180}\pi \tag{4.24}$$

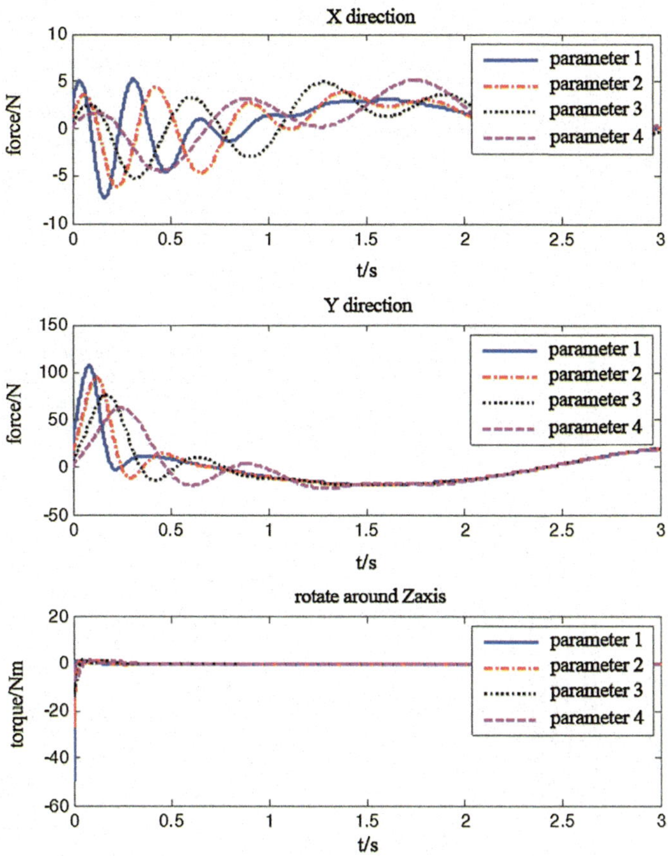

Fig. 4.10 Human-machine acting force in operation space when parameters of the human-machine interaction model change

$$q_{h3} = \frac{11}{10}\sin(t + \pi) \tag{4.25}$$

$\boldsymbol{q}_h = [q_{h1} \quad q_{h2} \quad q_{h3}]^T$ shows the trajectory of the human body, among which q_{h1} represents the ankle joint, q_{h2} represents the knee joint, and q_{h3} represents the hip joint. When the simulation time is 3 s, the movement trajectory shows that the body returns back to the original state from the initial state of squatting, and the joint angle of the initial state is $\boldsymbol{q}_h = \left[-\frac{5}{180}\pi \quad \frac{5}{180}\pi \quad 0\right]^T$. Set human-machine interaction model parameters $\boldsymbol{K}_{Pf} = \text{diag}[10{,}000 \quad 75{,}000 \quad 500]$, $\boldsymbol{K}_{Df} = \text{diag}[200 \quad 1000 \quad 10]$, and the controller parameters $\boldsymbol{K}_f = \text{diag}[0.9 \quad 0.9 \quad 0.9]$, $\boldsymbol{K}_D = \text{diag}[0.01 \quad 0.01 \quad 0.001]$.

The simulation results are as shown from Figs. 4.4, 4.5, 4.6, 4.7, 4.8 and 4.9. Figure 4.4 shows the trajectory tracking curve, illustrating the exoskeleton suit has

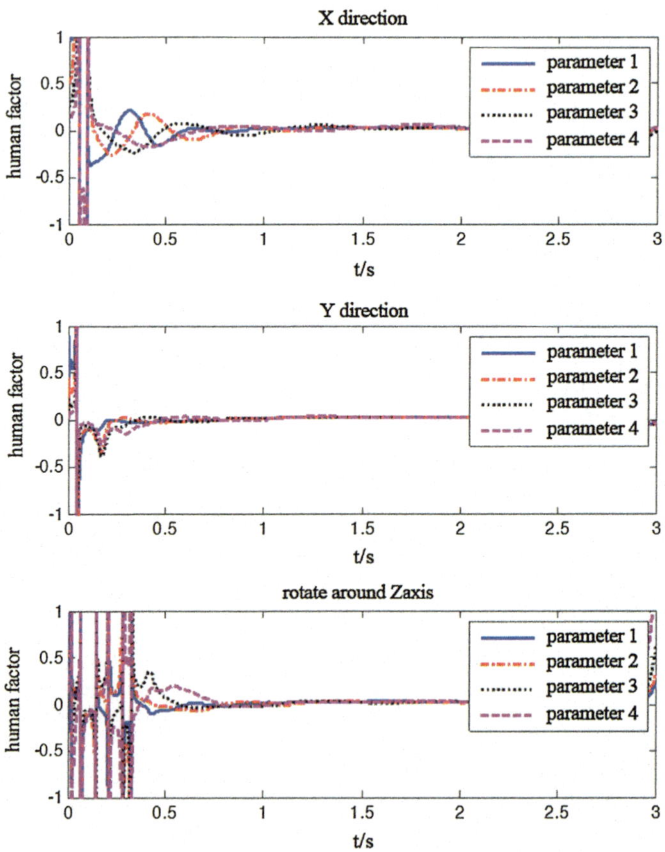

Fig. 4.11 Human factor in operation space when parameters of the human-machine interaction model change

succeeded in tracking the movement trajectory of the human body with good effect. Figure 4.5 shows the trajectory tracking error and the error is very small, which is mainly at 10^{-3} orders of magnitude but not zero.

Figure 4.6 shows the human-machine interaction force in the operation space, among which solid lines show the human-machine force when the actuator does not work, namely all the control force of the exoskeleton suit is exerted by the operator and dotted lines show the human-machine force when the actuator is at work, when human-machine force becomes very small, but not zero, and most of the control force is exerted by the actuator. Figure 4.7 shows the human factor ***HF*** in the operation space, from which we can see that each element of ***HF*** is mainly within ± 0.1, and the maximum human factor $\boldsymbol{HFM} = [0.0453 \quad 0.2934 \quad 0.2872]$.

Figure 4.8 shows that after Jacobian transformation, the force of the operation space gets the force moment of the joint space. Similarly, the solid line shows human-machine force moment in joint space when the actuator does not work, and

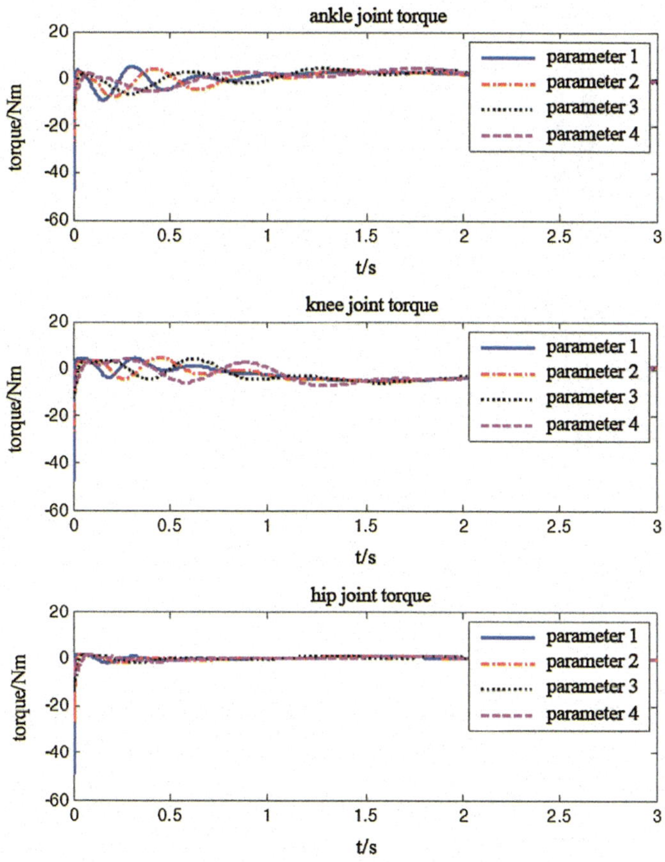

Fig. 4.12 Human-machine acting force in joint space when parameters of the human-machine interaction model change

the dotted line shows human-machine force moment in joint space when the actuator is at work. Figure 4.9 shows the human factor ***HF*** in joint space, and similarly, each element of ***HF*** is mainly within ±0.1, and the maximum human factor ***HFM*** $= [0.0814 \quad 0.0907 \quad 0.1442]$. From Figs. 4.6, 4.7, 4.8 and 4.9, we can see that when the actuator is working, the human-machine acting force/torque is greatly decreased, accounting for about 10% of the force needed by the exoskeleton suit movement, and so the operator's energy consumption is greatly reduced, which

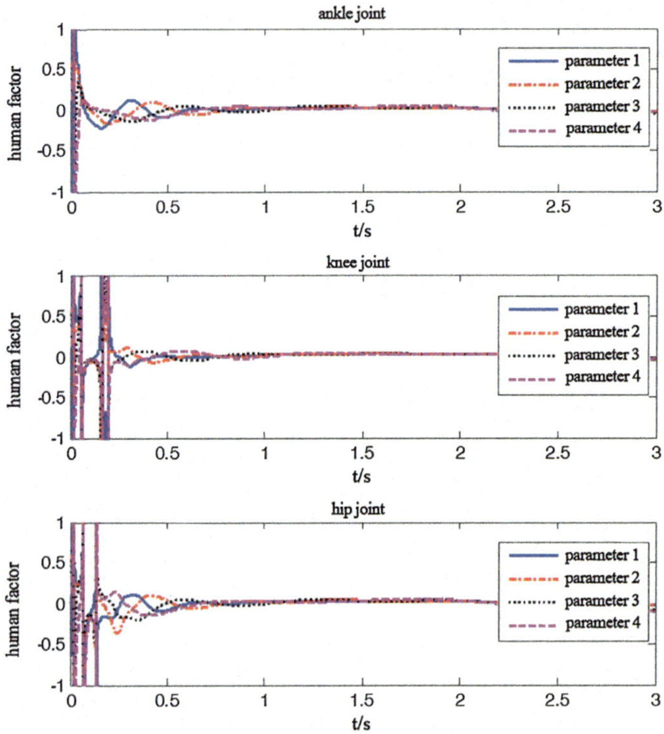

Fig. 4.13 Human factor in joint space when parameters of the human-machine interaction model change

Table 4.2 Maximum human factor when human-machine interaction model parameters change

Parameters	*X* direction	*Y* direction	Rotate around *Z* axis	Ankle joint	Knee joint	Hip joint
Parameter 1	0.0771	0.1120	0.1216	0.0963	0.0341	0.0687
Parameter 2	0.0652	0.0994	0.1219	0.0806	0.0444	0.2950
Parameter 3	0.0552	0.0804	0.1509	0.0925	0.0584	0.3626
Parameter 4	0.0538	0.0656	0.1714	0.0705	0.0487	0.2723

shows the designed controller has reached the desired control purpose, effective, and feasible.

4.4.1 Human-Machine Interaction Model Parameter's Influence to Force

The human in the human-machine interaction model in the above simulation also can be seen as the environment, whose parameter is unknown; just as different people wear exoskeleton suit, their strength is different, and the human-machine interaction model parameter reflects the differences. In order to verify the effect of different people wearing exoskeleton suit, the following simulation has been made, and let

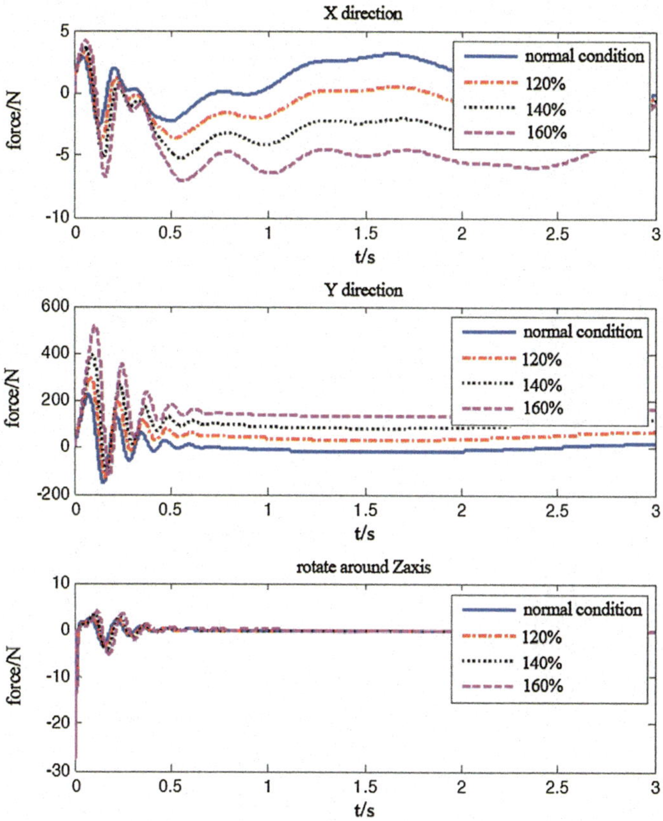

Fig. 4.14 Human-machine acting force in operation space under load change

$$\boldsymbol{K}_{\mathrm{Pf}} = \mathrm{diag}[\, K_{\mathrm{pfx}} \quad K_{\mathrm{pfy}} \quad K_{\mathrm{pfrz}} \,] \tag{4.26}$$

$$\boldsymbol{K}_{\mathrm{Df}} = \mathrm{diag}[\, K_{\mathrm{dfx}} \quad K_{\mathrm{dfy}} \quad K_{\mathrm{dfrz}} \,] \tag{4.27}$$

Here, select the following four different sets of parameters to simulate, as shown in Table 4.1.

Human-machine interaction model parameters become smaller, representing the decrease of the human strength, or the physical weakening. Figure 4.10 shows the contrast curve of the human-machine interaction force in the operation space; from the human-machine interaction force curve of its *X* direction and *Y* direction, we can see that when the human-machine interaction model parameters are becoming smaller, the time span of human-machine acting force forming the first wave crest (valley) becomes longer, but the amplitude becomes smaller, and the oscillation time becomes longer, which represents that the speed of human action becomes

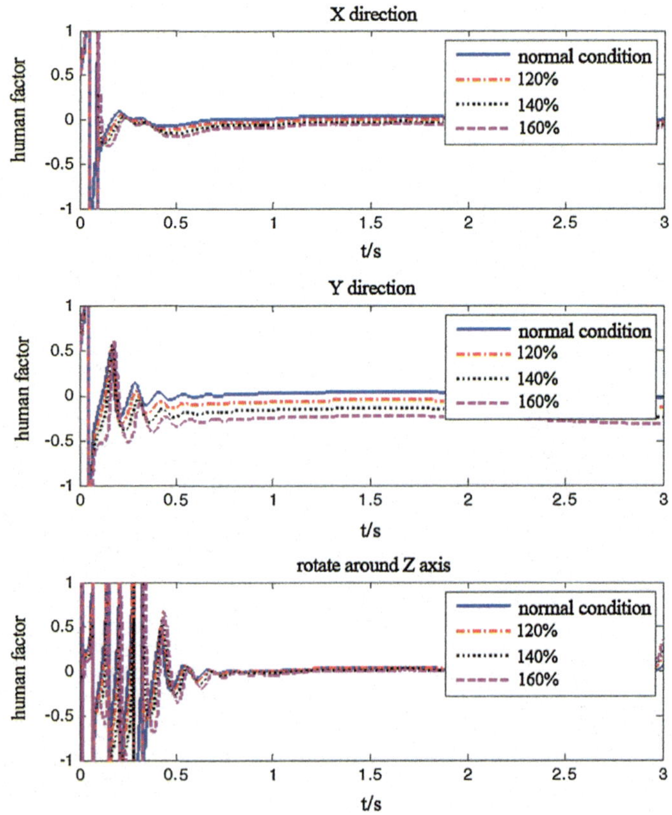

Fig. 4.15 Human factor in operation space in the case of load change

slower, and the action's time becomes longer, conforming to the actual situation. Figure 4.10 also shows the designed controller remains stable for different people and makes the human-machine acting force to decrease. Figure 4.11 shows the human factor ***HF*** in operation space, from which we can see that the human factor mostly remains within ±0.1, illustrating that human-machine interaction model has no big effect on human factor. Figures 4.12 and 4.13 illustrate the same problem from the viewpoint of joint space.

Table 4.2 gives the maximum human factor in the operation space and joint space when the human-machine interaction model parameters change, from which we can see that most elements' parameters are within 0.18, and only that of the hip joint is slightly bigger, because the human-machine acting force of the hip joint itself is not large.

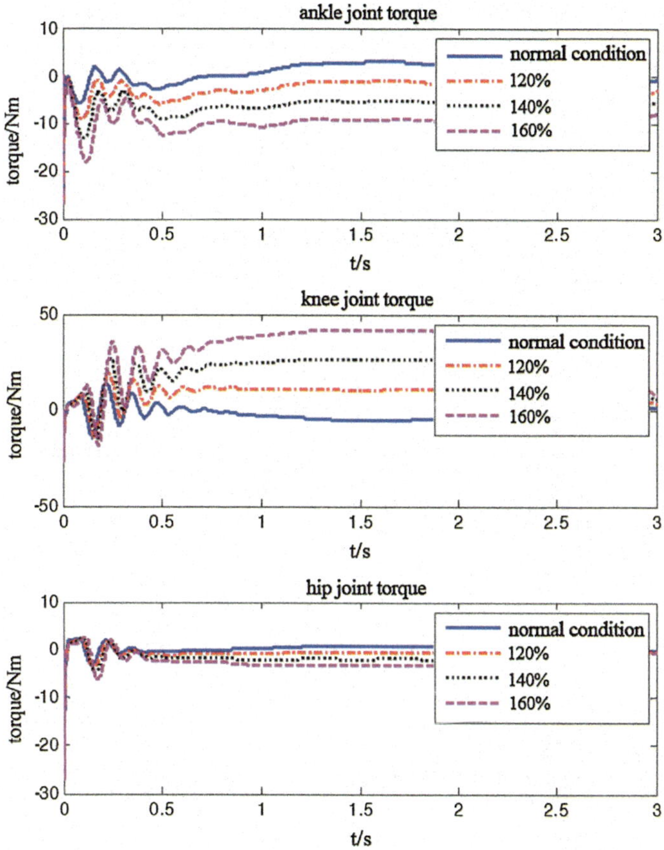

Fig. 4.16 Human-machine acting force in joint space in the case of load change

4.4.2 Load Varying's Influence to Force

Gravity compensation term in the controller includes several rigid mass attributes of the exoskeleton suit, among which the exoskeleton suit's trunk mass includes two aspects: One is the mass of the trunk itself of the exoskeleton suit, and the other is the mass of what is being loaded. Because in the process of the actual use, the change of the loading will cause the change of trunk weight, and this change is unknown, therefore, the human-machine force will be influenced, and need to make the simulation research.

Assume that the mass parameter of the exoskeleton suit's trunk taken from the controller is the weight of the exoskeleton suit's trunk itself, namely the weight without load. Make simulation for such four situations, respectively: (1) the normal situation, namely without load; (2) 120%, namely the total mass of the load and the exoskeleton suit accounts for 120% of the mass of the original exoskeleton suit's trunk; (3) 140%, similar to (2); (4) 160%, similar to (2) and (3).

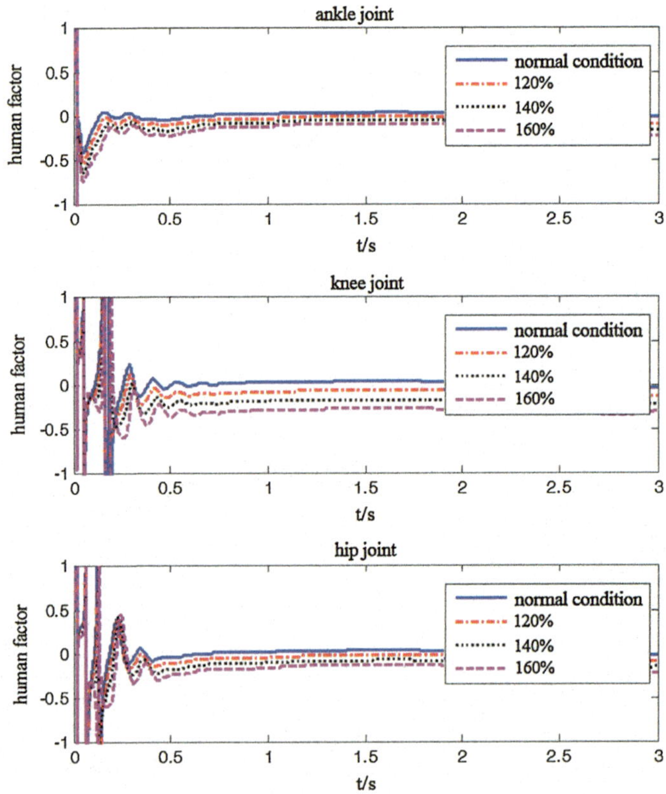

Fig. 4.17 Human factor in joint space in the case of load change

When simulating, take the second group of parameters of Table 4.1 as the human-machine interaction model parameters. The weight of the trunk itself without load is 47.488 kg.

The simulation results are as shown from Figs. 4.14, 4.15, 4.16, 4.17 and 4.18. Because all the trajectory tracking effect of the exoskeleton suit in the four cases is very good, and the tracking error is very small, it is no longer to give the simulation diagram. Figure 4.14 shows the human-machine acting force contrast curve in the operation space under the condition of load change, from which we can see that due to the increase of load, the human-machine force is also increasing especially in the Y direction, illustrating the robustness of the designed controller is poor for the load variation. Figure 4.15 shows the human factor $\boldsymbol{HF}$ in the operation space under the condition of load change, from which we can see that due to the change of the load, $\boldsymbol{HF}$ is no longer toward zero, but maintaining at a constant value. For example, for

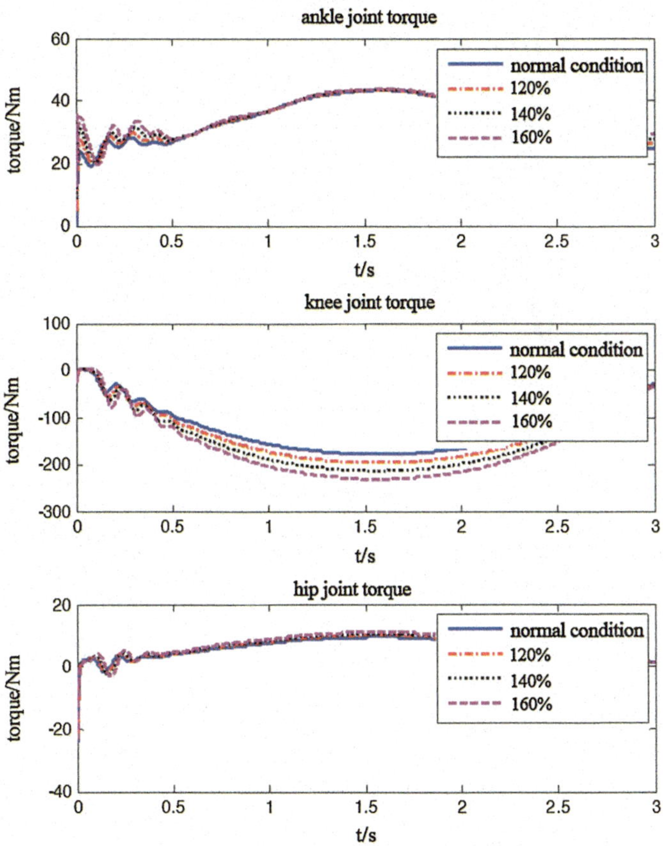

Fig. 4.18 Torque exerted by the actuator in joint space in the case of load change

Table 4.3 Maximum human factor when the load changes

Parameter	*X* direction	*Y* direction	Rotate around *Z* axis	Ankle joint	Knee joint	Hip joint
Normal circumstances	0.0652	0.0994	0.1219	0.0806	0.0444	0.2950
120%	0.0396	0.3122	0.3947	0.0949	0.1305	0.3002
140%	0.0554	0.4119	0.4727	0.1368	0.1771	0.3070
160%	0.0752	0.5395	0.5916	0.1878	0.2769	0.3138

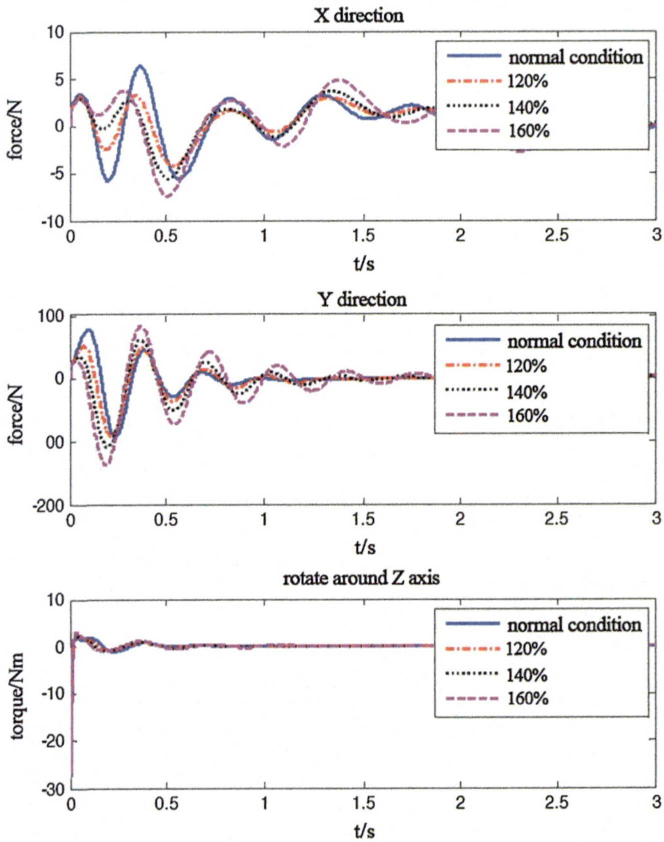

Fig. 4.19 Human-machine acting force in operation space in the case of load change under PI control

the fourth case, the changes of the ***HF*** are obvious in the *X* direction and *Y* direction, and especially in the *Y* direction, they always remain at around 0.3, illustrating that in this movement section, the acting force exerted by human accounts for about 30% of the resultant external force. Figure 4.16 shows the contrast curve of the human-machine acting force torque in joint space; Fig. 4.17 shows the human factor in joint space. Relative to Figs. 4.14 and 4.15, these two figures in joint space also illustrate the changes of the human-machine acting force torque are also affected by the load changes. Figure 4.18 shows the torque exerted by the actuator when load is changing, from which we can see that with the increase of load, the torque increased by the actuator is limited, not enough to compensate for the needed torque due to load increase, thus causing that human need exert extra torque to complement. Table 4.3 shows the maximum human factor ***HFM*** when the load is changing, from which we can see that when the load

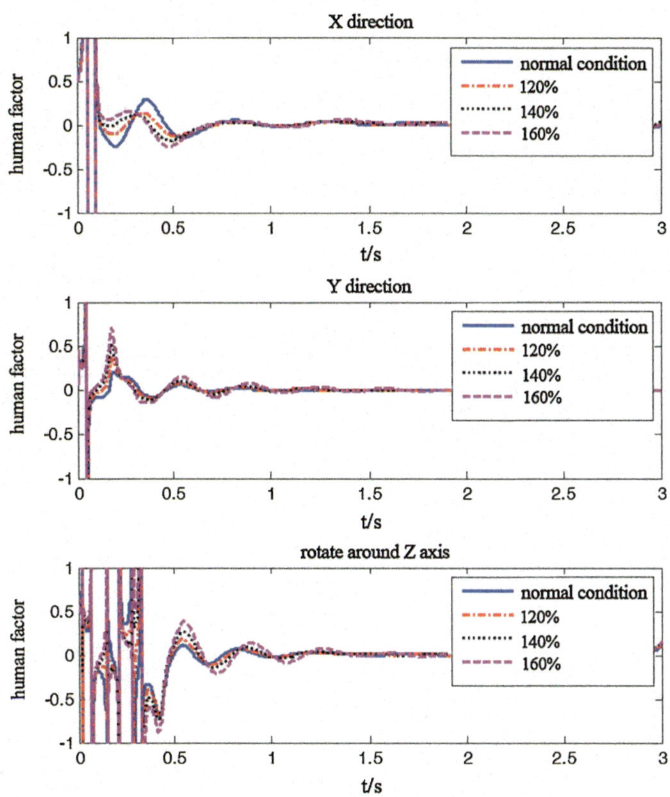

Fig. 4.20 Human factor in operation space in the case of load change under PI control

increases, except in the X direction, the ***HFM*** in other directions and in joint space has apparently increased.

4.4.3 Load Varying's Influence to Force Under PI Control

In order to overcome the influence of load change on human-machine force under PI control in Sect. 4.4.2, we introduce the proportional integral control, namely the PI control, and let

$$\tau = \boldsymbol{K}_{\mathrm{f}}\boldsymbol{f} + \boldsymbol{K}_{\mathrm{I}}\int_{0}^{t}\boldsymbol{f}\mathrm{d}t \tag{4.28}$$

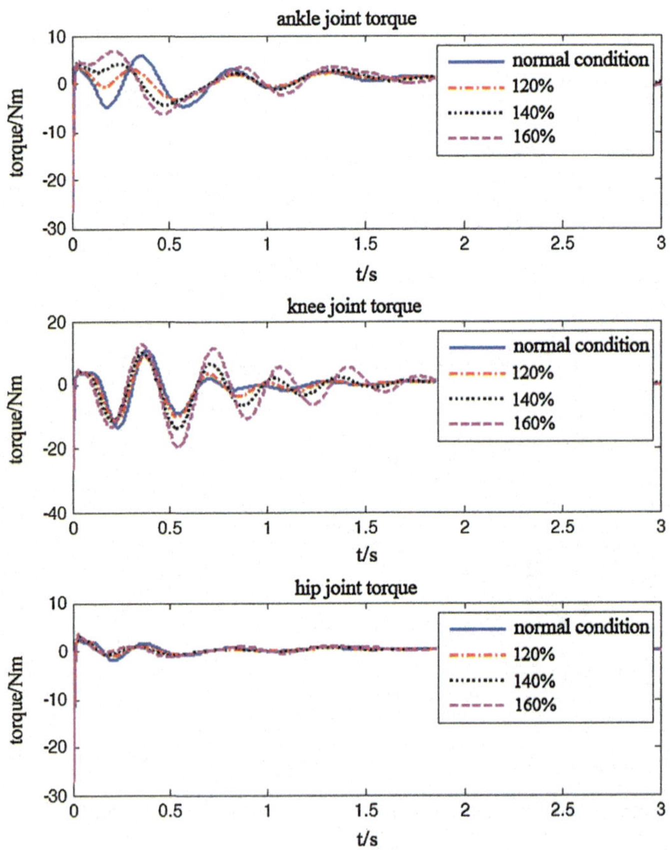

Fig. 4.21 Human-machine acting force in joint space in the case of load change under PI control

Introducing the integral is to make up for additional human-machine acting force exerted by the operator when the load changes. Similarly, simulate the four cases as such described in the Sect. 4.4.2. When simulating, take the second group of parameters from Table 4.1 as the proportion coefficient matrix $\boldsymbol{K}_{\mathrm{I}} = \mathrm{diag}[\,1 \quad 20 \quad 10\,]$ and the human-machine interaction model parameters. Other parameters are the same as the simulation parameters in the Sect. 4.4.

The simulation results are as shown from Fig. 4.19, 4.20, 4.21, 4.22, and 4.23. Figure 4.19 is corresponding to Fig. 4.14 and shows the human-machine force of operation space. Figure 4.21 is corresponding to Fig. 4.16 and shows human-machine force of joint space. From the figures, we can see that the introduction of the integral element effectively restrains the influence of load change on human-machine force; when the load increases, the smaller human-machine acting force is still remained, especially obvious in the *X* direction, *Y* direction of the operation space and ankle joint, knee joint of joint space, making the robustness of controller for the load variation to be enhanced greatly. Figure 4.20 is

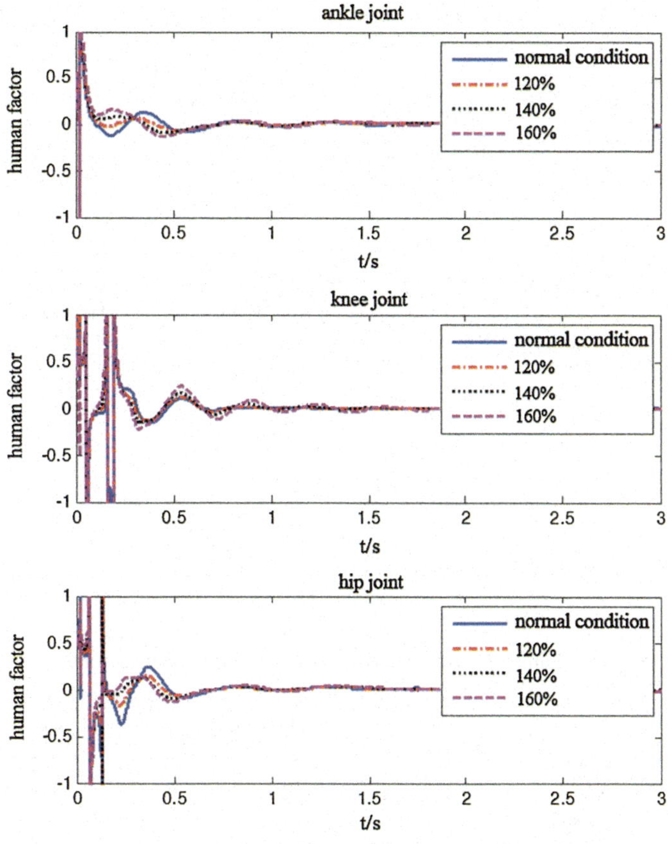

Fig. 4.22 Human factor in joint space in the case of load change under PI control

corresponding to Fig. 4.15 and shows the human factor ***HF*** in operation space. Figure 4.22 is corresponding to Fig. 4.17 and shows the human factor ***HF*** in joint space. By contrasting them, we can see that after adding PI control, ***HF*** converges near zero fast, which shows that the influence of the load change on the acting force exerted by human is very small. Figure 4.23 shows the torque curve exerted by the

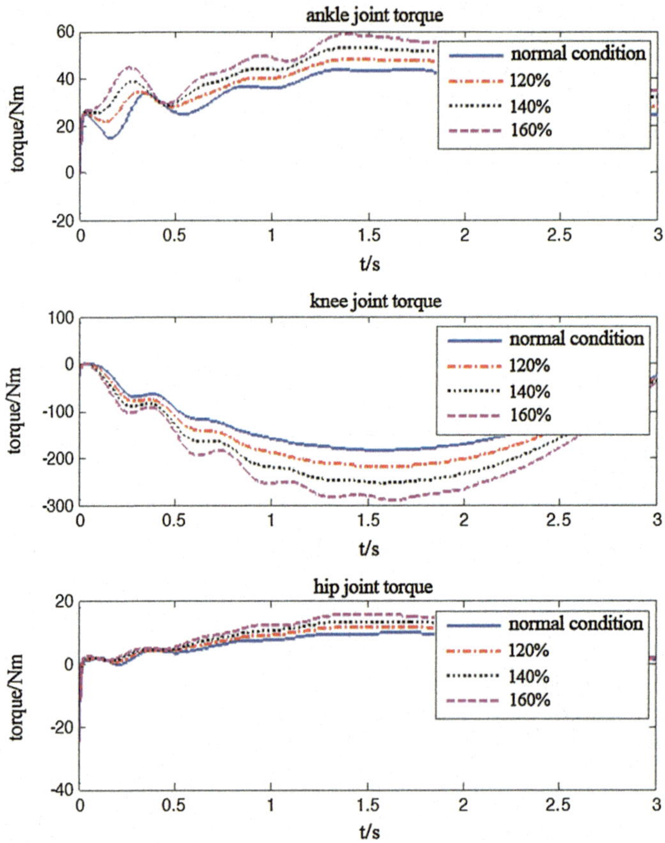

Fig. 4.23 Torque exerted by the actuator in joint space in the case of load change under PI control

Table 4.4 Maximum human factor in the case of load change under PI control

Parameters	*X* direction	*Y* direction	Rotate around *Z* axis	Ankle joint	Knee joint	Hip joint
Parameter 1	0.0670	0.0978	0.1884	0.0660	0.0882	0.2765
Parameter 2	0.0454	0.0966	0.1990	0.0641	0.0738	0.2694
Parameter 3	0.0582	0.1146	0.2308	0.0621	0.0903	0.2623
Parameter 4	0.0784	0.1442	0.2593	0.0708	0.1288	0.2554

actuator, from which we can see that with the increase of load, torque exerted by the actuator increases gradually, effectively compensating the impact caused by the load increase.

Table 4.4 shows the maximum human factor ***HFM*** in the case of the load change under PI control. Contrasted with Table 4.3, we can see that the change of the ***HFM*** is very small.

References

1. Wen Shuhuan. Intelligent Control and Trajectory Tracking Experiment Research of Uncertain Robot Force Position [D]. Yanshan University, Doctoral Dissertation, Qin Huangdao: 2004.
2. Liu Yuping. The Establishment and Research of Robot's Force Control Simulation Platform [D]. Shanghai University, Master's Thesis, Shanghai: 2007.
3. Zhang Guangli. Force Control System and Advanced Control Strategy Research for Open Structure Robot [D]. Shanghai Jiaotong University, Doctoral Dissertation, Shanghai: 2003.
4. Zhang Zhengwei. The Force Control Research of Robot End Effector [D]. Suzhou University, Master's Thesis, Suzhou: 2005.
5. Volpe, Richard, Pradeep Khosla. A Theoretical and Experimental Investigation of Explicit Force Control Strategies for Manipulators [J]. IEEE Transactions on Automatic Control, 1993, 38 (11): 1634–1650.
6. Eppinger, S.D., Seering W.P. Understanding Bandwidth Limitations in Robot Force Control [C]. Proceeding of the IEEE International Conference on Robotics and Automation, 1987: 904–909.
7. Whitney, D.E. Historical Perspective and State of the Art in Robot Force Control [J]. The International Journal of Robotics Research, 1987, 6 (1): 3–14.
8. Murat, Cenk Cavusoglu, Joseph Yan, S. Shankar Sastry. A Hybrid System Approach to Contact Stability and Force Control in Robotic Manipulators [C]. In Proceedings of the 12th IEEE International Symposium on Intelligent Control (ISIC'97), Istanbul, Turkey, 1997: 143–148.
9. Yin Yuehong, Wei Zhongxin, Huang Xiaoxi. Force Sensing and Force Control Technology for Intelligent Machine System [M]. Beijing: National Defence Industry Press, 2002.
10. Siciliano, Bruno, Luigi Villani. Robot Force Control [M]. Boston/Dordrecht/London: Kluwer Academic Publishers, 1999.
11. Winter, D A. Biomechanics of Human Movement [M]. New York: John Wiley and Sons, 1979.

Chapter 5
Force Control of Exoskeleton Suit Based on Inner Position Loop

5.1 Introduction

The robot's explicit force control can be divided into two kinds: the explicit force control based on the position and the explicit force control based on the force [1, 2]. The structure of the explicit force control based on the force is same as that of the direct force control described in the previous chapter, and this chapter will introduce the explicit force control method based on the position. The structure of the explicit force control based on the position is same as that of the damping control, whose characteristics are to determine a admittance matrix according to the force and position of the operation task; the admittance matrix determines the relationship between the contact force error and the velocity disturbance, achieving the desired contact force by controlling the operating arm movement with the position control subsystem. In the damping control, force/position transformation is realized by admittance, and its basic idea is to use the position control subsystem (inner position loop) as the basic system and to regulate the system admittance to complete a given smooth task [1, 3]; therefore, it is a kind of indirect force control method.

Now, it analyzes the force control of the exoskeleton suit in accordance with the above explicit force control concept based on the position. The force of human acting on the exoskeleton suit is closely linked with position error between the human and the exoskeleton suit; if the position deviation between the exoskeleton suit and human exists, the acting force between them will be produced; therefore, by controlling the position of the exoskeleton suit, we can control the acting force between them indirectly, and the idea is consistent with that of the explicit force control. In the explicit force control, however, generally, the reference position signal of the inner position loop has been known, and generally, we get the position correction signal through the force controller; for example, in the damping control, after transforming the force/position through the admittance matrix, what we get is the position correction. And in the exoskeleton suit system, the exoskeleton suit's reference position is the movement position of human body. As stated earlier, using

Z. Yang et al., *Force Control Theory and Method of Human Load Carrying Exoskeleton Suit*, DOI 10.1007/978-3-662-54144-9_5

sensors to measure movement position of the human body needs to install complex sensors on the human body, thus leading to the inconvenience of putting on and taking off the exoskeleton suit, hindering the body movement, and reducing the human body's comfort degree; therefore, measuring the position of the human body directly and using it as a reference position signal of the exoskeleton suit is not suitable. We can use the force sensor to measure the force exerted on the exoskeleton suit by human and, according to the force, estimate the reference position of the exoskeleton suit to complete the force control of the exoskeleton suit indirectly.

The force control method of the exoskeleton suit based on the inner position loop described in this chapter is to design the position control loop of the exoskeleton suit first to ensure that the exoskeleton suit can move precisely according to the reference trajectory and then to generate the reference trajectory of the inner position loop in operation space through the feedback information of multi-dimensional force/torque sensor to control the exoskeleton suit tracking the human's movement. This chapter introduces two kinds of inner position loop controllers based on the static and dynamic models and simulation research for the two inner position loops is carried out, respectively. At the same time, it introduces the force controller based on the inner position loop, analyzes the stability of the designed controller, simulates the support leg's squatting and rising action of the exoskeleton suit, verifies the feasibility and effectiveness of the controller, and analyzes the robustness of the controller.

5.2 Inner Position Loop

For the force control based on the inner position loop, first, a stable inner position loop and, then, the force control loop are designed. Therefore, the position control inner loop is studied first. Inner position loop refers to that through giving the desired joint angle signal of the exoskeleton suit, the joint motor of the exoskeleton suit to rotate to the desired angle is controlled. And if from the point view of operation space, the inner position loop refers to that through giving the desired orientation signal of the end effector of the exoskeleton suit (the center of gravity of the exoskeleton suit's trunk under supporting condition, the center of gravity of the exoskeleton suit's feet under swing condition) control joint motor rotation of the exoskeleton suit, enabling the exoskeleton suit's terminal to move to the desired position. The inner position loop described in this chapter can be divided into two types: the inner position loop based on the static model and the inner position loop based on the dynamic model. The so-called inner position loop based on the static model refers to that when designing the controller of the position control loop, only the system's static gravity term is considered, without considering the dynamic term, and correspondingly, the inner position loop based on the dynamic model refers to that when designing the controller, only the dynamic term is considered.

5.2.1 Inner Position Loop Based on the Static Model

If we take no account of human-machine interaction and other disturbance torque, by Eq. (2.74), we can get the kinetic model of the exoskeleton suit

$$\boldsymbol{H}(\boldsymbol{q})\ddot{\boldsymbol{q}} + \boldsymbol{C}(\boldsymbol{q}, \dot{\boldsymbol{q}})\dot{\boldsymbol{q}} + \boldsymbol{F}\dot{\boldsymbol{q}} + \boldsymbol{G}(\boldsymbol{q}) = \boldsymbol{T}_{\mathrm{a}} \tag{5.1}$$

Among them, for the meaning of each matrix, please refer to the Sect. 4.2.

Through designing the appropriate control activity, the control force and torque $\boldsymbol{T}_{\mathrm{a}}$ of the exoskeleton suit's terminal can be formed to control the terminal of the exoskeleton suit to reach the desired position and orientation in the operation space, as shown in Fig. 5.1.

If the controller is designed based on the model's static characteristics, there is

$$\boldsymbol{T}_{\mathrm{a}} = \boldsymbol{J}^{\mathrm{T}}(\boldsymbol{q})\boldsymbol{\tau} - \boldsymbol{K}_{\mathrm{D}}\dot{\boldsymbol{q}} + \boldsymbol{G}(\boldsymbol{q}) \tag{5.2}$$

where $\boldsymbol{\tau}$ represents the generalized force and torque exerted by the actuator; $\boldsymbol{G}(\boldsymbol{q})$ is used to compensate the gravity torque; $\boldsymbol{K}_{\mathrm{D}}$ is a positive definite gain matrix; and $-\boldsymbol{K}_{\mathrm{D}}\dot{\boldsymbol{q}}$ provides an additional damping torque for the joint to improve the dynamic response process of the system. The control based on Eq. (5.2) is the so-called ratio differential control with gravity compensation [4].

Then, we design the middle controlled quantity $\boldsymbol{\tau}$ as

$$\boldsymbol{\tau} = \boldsymbol{K}_{\mathrm{P1}}(\boldsymbol{x}_{\mathrm{c}} - \boldsymbol{x}_{\mathrm{e}}) \tag{5.3}$$

where $\boldsymbol{x}_{\mathrm{c}}$ and $\boldsymbol{x}_{\mathrm{e}}$ represent the reference position and the actual position of the exoskeleton suit's terminal in operation space, respectively, and $\boldsymbol{K}_{\mathrm{P1}}$ is the proportional coefficient that can be designed. Then, the system forms a position control closed loop.

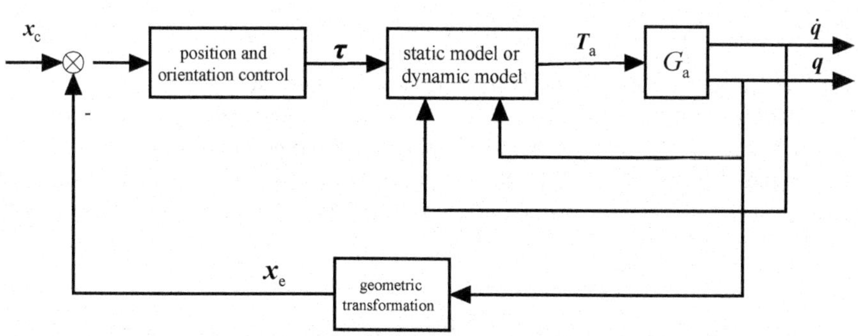

Fig. 5.1 Position control principle diagram of the exoskeleton suit

After substituting Eqs. (5.3) and (5.2) into Eq. (5.1), we can get the closed-loop dynamic state of the system:

$$\boldsymbol{H}(\boldsymbol{q})\ddot{\boldsymbol{q}} + \boldsymbol{C}(\boldsymbol{q},\dot{\boldsymbol{q}})\dot{\boldsymbol{q}} + (\boldsymbol{F} + \boldsymbol{K}_{\mathrm{D}})\dot{\boldsymbol{q}} = \boldsymbol{J}^{\mathrm{T}}(\boldsymbol{q})\boldsymbol{K}_{\mathrm{P1}}(\boldsymbol{x}_{\mathrm{c}} - \boldsymbol{x}_{\mathrm{e}}) \tag{5.4}$$

In order to analyze the dynamic stability, let $\Delta\boldsymbol{x}_{\mathrm{ce}} = \boldsymbol{x}_{\mathrm{c}} - \boldsymbol{x}_{\mathrm{e}}$, and introduce the following Lyapunov function

$$V = \frac{1}{2}\dot{\boldsymbol{q}}^{\mathrm{T}}\boldsymbol{H}(\boldsymbol{q})\dot{\boldsymbol{q}} + \frac{1}{2}\Delta\boldsymbol{x}_{\mathrm{ce}}{}^{\mathrm{T}}\boldsymbol{K}_{\mathrm{P1}}\Delta\boldsymbol{x}_{\mathrm{ce}} \tag{5.5}$$

Then, $V > 0$, and differentiating V, we obtain

$$\begin{aligned}\dot{V} &= \frac{1}{2}\dot{\boldsymbol{q}}^{\mathrm{T}}\dot{\boldsymbol{H}}(\boldsymbol{q},\dot{\boldsymbol{q}})\dot{\boldsymbol{q}} + \dot{\boldsymbol{q}}^{\mathrm{T}}\boldsymbol{H}(\boldsymbol{q})\ddot{\boldsymbol{q}} + \Delta\dot{\boldsymbol{x}}_{\mathrm{ce}}{}^{\mathrm{T}}\boldsymbol{K}_{\mathrm{P1}}\Delta\boldsymbol{x}_{\mathrm{ce}} \\ &= -\dot{\boldsymbol{q}}^{\mathrm{T}}(\boldsymbol{F} + \boldsymbol{K}_{\mathrm{D}})\dot{\boldsymbol{q}}\end{aligned} \tag{5.6}$$

Because both the matrix $\boldsymbol{F}$ and $\boldsymbol{K}_{\mathrm{D}}$ are positive definite, and $\dot{V}$ is not, the system is asymptotically stable in the balance point. So the terminal of the exoskeleton suit has achieved the desired orientation adjustment control.

5.2.2 Inner Position Loop Based on the Dynamic Model

If the controller is designed based on the model's dynamic characteristics, then there is

$$\boldsymbol{T}_{\mathrm{a}} = \boldsymbol{H}(\boldsymbol{q})\boldsymbol{\tau} + \boldsymbol{C}(\boldsymbol{q},\dot{\boldsymbol{q}})\dot{\boldsymbol{q}} + \boldsymbol{F}\dot{\boldsymbol{q}} + \boldsymbol{G}(\boldsymbol{q}) \tag{5.7}$$

Substituting Eq. (5.7) into Eq. (5.1), and assuming that the $\boldsymbol{H}(\boldsymbol{q})$ is always nonsingular, then, we obtain the dynamic state of the system as follows:

$$\ddot{\boldsymbol{q}} = \boldsymbol{\tau} \tag{5.8}$$

The system turns into a linear decoupling double integral system.

In fact, only through compensating all parameter items of Eq. (5.1) accurately, Eq. (5.8) can be attained. So we need to obtain the precise dynamic model of the system through identifying the dynamic parameters of Eq. (5.1). When it cannot be compensated accurately, a disturbance term $\boldsymbol{\delta}$ will be produced in Eq. (5.8), namely

$$\ddot{\boldsymbol{q}} = \boldsymbol{\tau} + \boldsymbol{\delta} \tag{5.9}$$

In fact, the disturbance term $\boldsymbol{\delta}$ is mainly caused by the unmodeled dynamics, such as friction torque.

Because Eq. (5.8) includes $\ddot{\boldsymbol{q}}$, we differentiate Eq. (2.45) and obtain

$$\ddot{\boldsymbol{x}}_{\mathrm{e}} = \boldsymbol{J}(\boldsymbol{q})\ddot{\boldsymbol{q}} + \dot{\boldsymbol{J}}(\boldsymbol{q},\dot{\boldsymbol{q}})\dot{\boldsymbol{q}} \tag{5.10}$$

Equation (5.10) gives the relationship between joint angular acceleration and the linear acceleration and angular acceleration of the exoskeleton suit's terminal. Therefore, the new controlled quantity is designed as follows:

$$\tau = \boldsymbol{J}(\boldsymbol{q})^{-1}\left[\boldsymbol{u} - \dot{\boldsymbol{J}}(\boldsymbol{q},\dot{\boldsymbol{q}})\dot{\boldsymbol{q}}\right] \tag{5.11}$$

where $\boldsymbol{u}$ is the new control variable formed in the operation space. Then, when taking no account of the disturbance term, through Eqs. (5.11), (5.10), and (5.8), we can obtain

$$\ddot{\boldsymbol{x}}_{\mathrm{e}} = \boldsymbol{u} \tag{5.12}$$

If we design $\boldsymbol{u}$ as

$$\boldsymbol{u} = \ddot{\boldsymbol{x}}_{\mathrm{c}} + \boldsymbol{K}_{\mathrm{D}}\Delta\dot{\boldsymbol{x}}_{\mathrm{ce}} + \boldsymbol{K}_{\mathrm{P1}}\Delta\boldsymbol{x}_{\mathrm{ce}} \tag{5.13}$$

after substituting Eq. (5.13) into Eq. (5.12), we can obtain the closed-loop dynamic state of the system as follows:

$$\Delta\ddot{\boldsymbol{x}}_{\mathrm{ce}} + \boldsymbol{K}_{\mathrm{D}}\Delta\dot{\boldsymbol{x}}_{\mathrm{ce}} + \boldsymbol{K}_{\mathrm{P1}}\Delta\boldsymbol{x}_{\mathrm{ce}} = 0 \tag{5.14}$$

For any selected positive definite matrices $\boldsymbol{K}_{\mathrm{D}}$ and $\boldsymbol{K}_{\mathrm{P1}}$, Eq. (5.14) is always exponentially stable, so the terminal of the exoskeleton suit has achieved the trajectory tracking for the $\boldsymbol{x}_{\mathrm{c}}$.

5.3 Force Control Based on the Inner Position Loop

Once the reference orientation signal of the exoskeleton suit is established, if it is a fixed value, the exoskeleton suit will keep a constant pose in operation space. At this time, if the operator exerts a disturbance force or disturbance torque on the terminal of the exoskeleton suit, it is hard to change the terminal's orientation, because the sensitivity of the inner position loop for the external disturbance is very small, unless the disturbance force or torque is big enough; of course, we do not expect that happens, because it will consume a huge amount of energy from the human body, and what we expect is that we can swing the exoskeleton suit with a very small force. An intuitive idea is to establish the relationship between the disturbance force/torque and the reference orientation of the exoskeleton suit. By disturbing the magnitude of the force/torque, we can estimate the magnitude of the reference orientation to control the exoskeleton suit's movement. Once the exoskeleton suit moves to the desired orientation of the operator, then disturbance force/torque exerted by the operator goes to zero; when the desired orientation will not change any longer, then the exoskeleton suit's tracking for the intention of human movement is achieved. Of course, in the process, we need to ensure that the disturbance force/torque exerted by the operator is as small as possible. Based on the thought, we will introduce a kind of force control method based on the inner position loop and describe in detail as follows:

Assume that there is following relationship between the reference orientation $\boldsymbol{x}_c$ of the exoskeleton suit's terminal and human-machine acting force $\boldsymbol{f}$, or between the reference orientation $\boldsymbol{x}_c$ of the exoskeleton suit's terminal and the generalized force of human exerting on the exoskeleton suit's terminal

$$\boldsymbol{x}_c = \boldsymbol{F}(\boldsymbol{f}) \tag{5.15}$$

And suppose the multi-dimensional force/torque sensor is used to measure the acting force $\boldsymbol{f}$; then, the force control principle diagram based on the inner position loop is shown in Fig. 5.2. What we need is to pay attention to the acting force $\boldsymbol{f}$ caused by the deviation between the operator's desired orientation and the orientation of the exoskeleton suit's terminal; not only $\boldsymbol{f}$ is used to estimate the desired orientation of the operator, but also it acts on the exoskeleton suit's terminal and is transformed to human-machine joint torque $\boldsymbol{T}_{hm}$ through the Jacobian inverse transformation, acting on the exoskeleton suit together with the control torque $\boldsymbol{T}_a$ exerted by the actuator.

To analyze the characteristics of the system, first, the kinetic model of the system is given. Because the human-machine interaction torque is added, the kinetic model of the exoskeleton suit becomes

$$\boldsymbol{H}(\boldsymbol{q})\ddot{\boldsymbol{q}} + \boldsymbol{C}(\boldsymbol{q},\dot{\boldsymbol{q}})\dot{\boldsymbol{q}} + \boldsymbol{F}\dot{\boldsymbol{q}} + \boldsymbol{G}(\boldsymbol{q}) = \boldsymbol{T}_a + \boldsymbol{T}_{hm} \tag{5.16}$$

If we design the inner position loop controller according to the static model as shown in Eqs. (5.2) and (5.3), and design

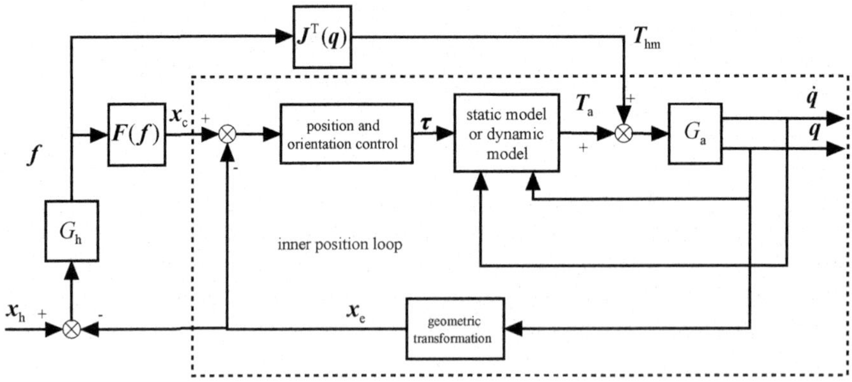

Fig. 5.2 Force control principle diagram based on the inner position loop

$$\boldsymbol{x}_{\mathrm{c}} = \boldsymbol{K}_{\mathrm{P1}}{}^{-1}\boldsymbol{K}_{\mathrm{P}}\boldsymbol{f} \tag{5.17}$$

then the force control outer loop is formed, in which $\boldsymbol{K}_{\mathrm{P}}$ is proportion coefficient that can be designed. After substituting Eqs. (5.2), (5.3), and (5.17) into Eq. (5.16), we can get the closed-loop dynamic state of the system

$$\boldsymbol{H}(\boldsymbol{q})\ddot{\boldsymbol{q}} + \boldsymbol{C}(\boldsymbol{q},\dot{\boldsymbol{q}})\dot{\boldsymbol{q}} + (\boldsymbol{F}+\boldsymbol{K}_{\mathrm{D}})\dot{\boldsymbol{q}} = \boldsymbol{J}^{\mathrm{T}}(\boldsymbol{q})(\boldsymbol{K}_{\mathrm{p}}+\boldsymbol{I})\boldsymbol{f} - \boldsymbol{J}^{\mathrm{T}}(\boldsymbol{q})\boldsymbol{K}_{\mathrm{P1}}\boldsymbol{x}_{\mathrm{e}} \tag{5.18}$$

Therefore, during the steady state,

$$\boldsymbol{f}_{\infty} = (\boldsymbol{K}_{\mathrm{P}}+\boldsymbol{I})^{-1}\boldsymbol{K}_{\mathrm{P1}}\boldsymbol{x}_{\mathrm{e}} \tag{5.19}$$

We can see that although we can design $\boldsymbol{K}_{\mathrm{P}}$ and $\boldsymbol{K}_{\mathrm{P1}}$ to make the human-machine acting force as small as possible, the steady-state error of the human-machine acting force still exists. In order to eliminate the steady-state error, we introduce the integral term, and let

$$\boldsymbol{x}_{\mathrm{c}} = \boldsymbol{K}_{\mathrm{P1}}{}^{-1}\left(\boldsymbol{K}_{\mathrm{P}}\boldsymbol{f} + \boldsymbol{K}_{\mathrm{I}}\int_0^t \boldsymbol{f}\mathrm{d}t\right) \tag{5.20}$$

where $\boldsymbol{K}_{\mathrm{I}}$ is the integral coefficient that can be designed. Then, we can ensure that in a steady state,

$$\boldsymbol{f}_{\infty} = 0 \tag{5.21}$$

By Eqs. (4.7), (4.8), and (4.9), we can know that whichever human-machine interaction model is used, when $\boldsymbol{f}_{\infty} = 0$, there is $\boldsymbol{x}_{\mathrm{e}} = \boldsymbol{x}_{\mathrm{h}}$.

By the same token, if we keep the inner position loop control law based on the dynamic model of Eqs. (5.7) and (5.13) changeless, Eq. (5.11) is redesigned as follows:

$$\boldsymbol{\tau} = \boldsymbol{J}(\boldsymbol{q})^{-1}\left[\boldsymbol{u} - \dot{\boldsymbol{J}}(\boldsymbol{q},\dot{\boldsymbol{q}})\dot{\boldsymbol{q}} - \boldsymbol{J}(\boldsymbol{q})\boldsymbol{H}^{-1}(\boldsymbol{q})\boldsymbol{J}^{\mathrm{T}}(\boldsymbol{q})\boldsymbol{f}\right] \tag{5.22}$$

Then after substituting Eqs. (5.7), (5.13), and (5.22) into Eq. (5.16), we can get the closed-loop dynamic of the system

$$\Delta\ddot{\boldsymbol{x}}_{\mathrm{ce}} + \boldsymbol{K}_{\mathrm{D}}\Delta\dot{\boldsymbol{x}}_{\mathrm{ce}} + \boldsymbol{K}_{\mathrm{P1}}\Delta\boldsymbol{x}_{\mathrm{ce}} = 0 \tag{5.23}$$

Similarly, during the steady state, there is

$$\boldsymbol{f}_{\infty} = \boldsymbol{K}_{\mathrm{P}}{}^{-1}\boldsymbol{K}_{\mathrm{P1}}\boldsymbol{x}_{\mathrm{e}} \tag{5.24}$$

So, the integral control term is also needed to be introduced to eliminate the steady-state error.

From the above analysis, it can be seen that the effect of the steady state generated by using dynamic model compensation and using static model compensation is same. But by comparison, we can see that the controller using the static model compensation is much simpler than that using the dynamic model compensation.

5.4 Simulation

5.4.1 Simulation of Inner Position Loop

For the force control method based on the inner position loop, first, a stable inner position loop need to be established, or the effect of force control cannot be guaranteed. Therefore, first, the control of the inner position loop is simulated to observe its control effect.

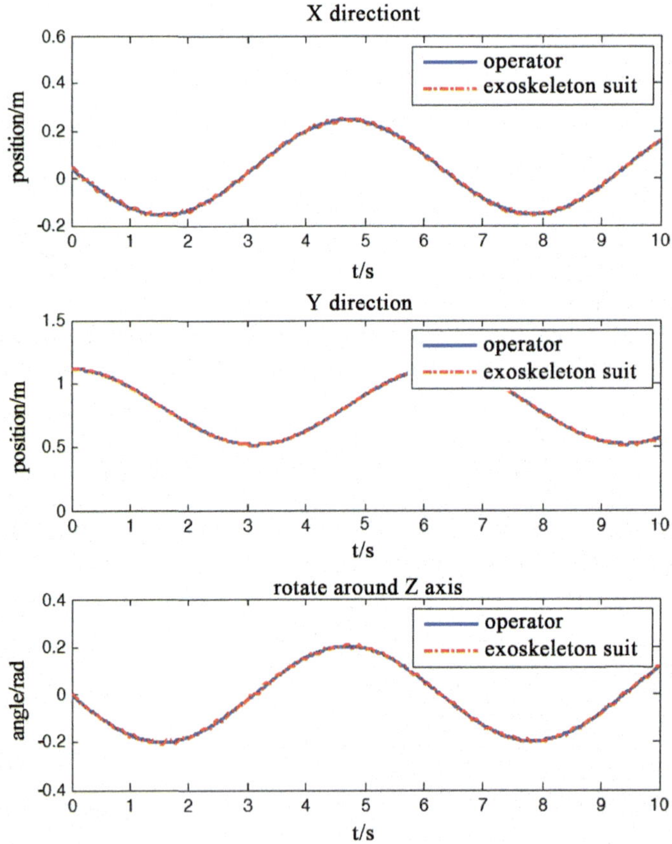

Fig. 5.3 Orientation tracking curve

The controlled object is the support leg model shown in Fig. 2.5a. When we design the inner position loop, no interaction between the exoskeleton suit and the operator is taken into account. Thus, the support leg can be regarded as an ankle fixed on the ground; shanks, thighs, and the trunk form a three-connecting-rod manipulator, and the trunk's center of gravity is regarded as its terminal. Similarly, SimMechanics toolbox of MATLAB is used to establish the mechanical model of the exoskeleton suit. For the mass and geometric attributes of the exoskeleton suit, we adopt the human body parameters of the Winter [5].

Using the ankle as the coordinate's null point and selecting the initial orientation of the exoskeleton suit's trunk, we obtain

$$\boldsymbol{x}_0 = [0.0435 \quad 1.114 \quad 0]^{\mathrm{T}} \tag{5.25}$$

Let the reference orientation be

$$\boldsymbol{x}_{\mathrm{c}} = [p_{\mathrm{x}} \quad p_{\mathrm{y}} \quad \theta]^{\mathrm{T}} + \boldsymbol{x}_0 \tag{5.26}$$

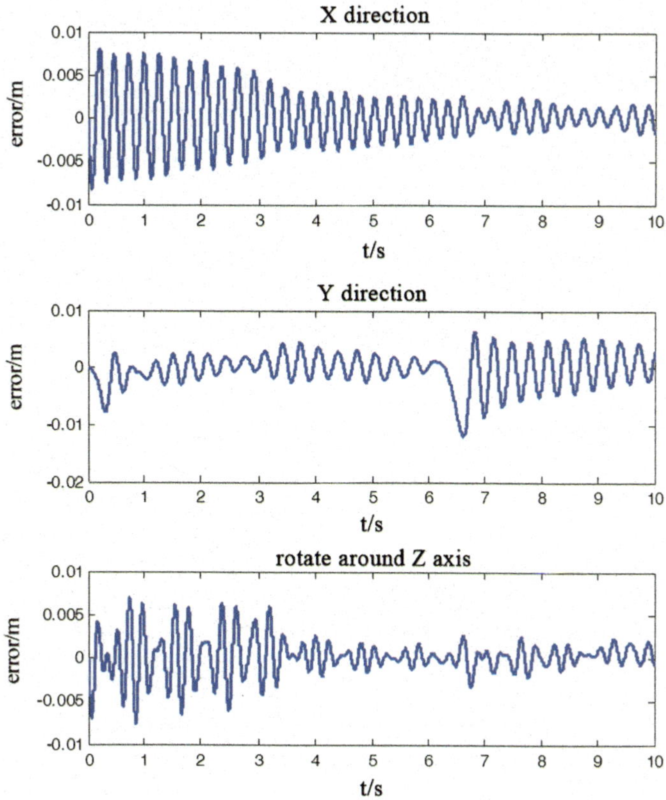

Fig. 5.4 Orientation tracking error curve

And select

$$p_x = -\frac{1}{5}\sin(t) \tag{5.27}$$

$$p_y = \frac{3}{10}\sin\left(t + \frac{\pi}{2}\right) - \frac{3}{10} \tag{5.28}$$

$$\theta = -\frac{1}{5}\sin(t) \tag{5.29}$$

5.4.1.1 The Inner Position Loop Based on the Static Model

The position controller based on the static model is shown in Eqs. (5.2) and (5.3). During simulation, the controller parameters are selected.

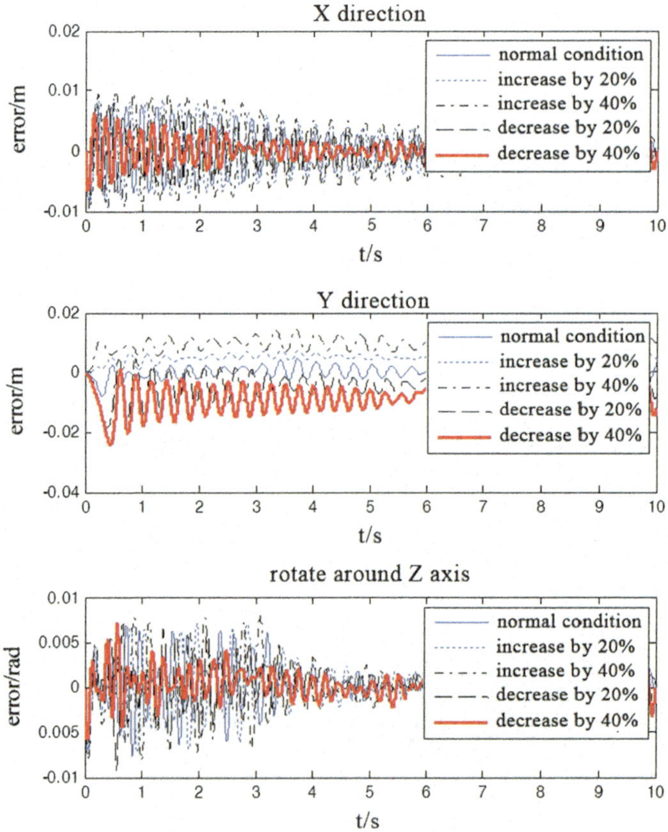

Fig. 5.5 Orientation tracking error contrast curve when load changes

$$\boldsymbol{K}_{\mathrm{D}} = \mathrm{diag}[\,0.01 \quad 0.01 \quad 0.001\,] \tag{5.30}$$

$$\boldsymbol{K}_{\mathrm{P1}} = \mathrm{diag}[\,30000 \quad 20000 \quad 300\,] \tag{5.31}$$

When the reference orientation, as shown in Eqs. (5.25) to (5.29), and the controller parameters, as shown in Eqs. (5.30) and (5.31), are given, the simulation results are shown in Figs. 5.3 and 5.4. Figure 5.3 shows that the effect of the exoskeleton suit tracking the reference orientation is very good. And Fig. 5.4 shows the tracking error curve, from which we can see that the error is small, at 10^{-2} orders of magnitude.

Because after the controller has been selected, when there is no self-adaptive mechanism, the controller parameters no longer change, and the load borne by the exoskeleton suit possibly changes, it simulates this situation to observe the control

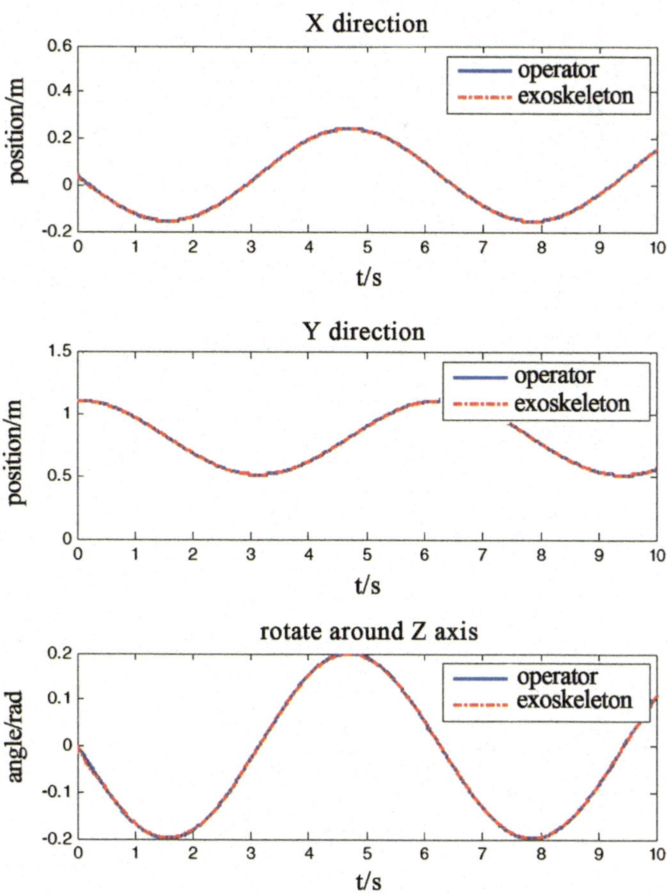

Fig. 5.6 Orientation tracking curve

effect of the inner position loop. Assume that the mass of the exoskeleton suit's trunk is the load, and simulate each of the five different conditions, respectively: The trunk's mass is in normal condition; the trunk's mass increases by 40 and 20% and reduces by 20 and 40%.

Figure 5.5 shows the orientation tracking error curve of five cases in three directions in operation space. In the five cases, both the position error and angle error are very small, illustrating that the effect of the designed position controller is good and the robustness for the load variation is very strong.

5.4.1.2 The Inner Position Loop Based on the Dynamic Model

The position controller based on the dynamic model is shown in Eqs. (5.7), (5.11), and (5.13). During simulation, the controller parameters are selected as follows:

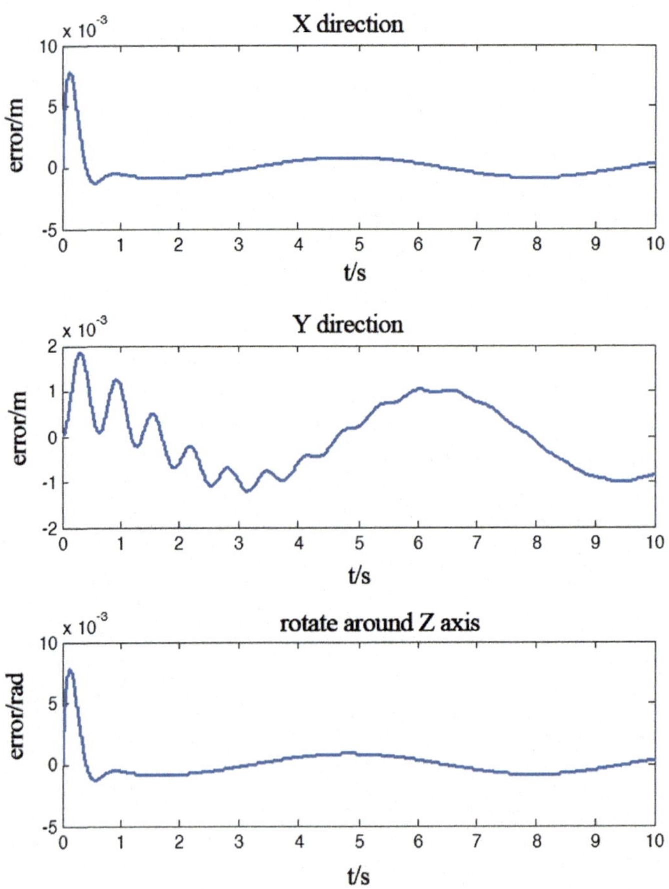

Fig. 5.7 Orientation tracking error curve

$$\boldsymbol{K}_{\mathrm{D}} = \mathrm{diag}[\,10 \quad 1 \quad 10\,] \tag{5.32}$$

$$\boldsymbol{K}_{\mathrm{P1}} = \mathrm{diag}[\,80 \quad 100 \quad 80\,] \tag{5.33}$$

When the reference orientation, as shown in Eqs. (5.25) to (5.29), and the controller parameters, as shown in Eqs. (5.32) and (5.33), are given, the simulation results are shown in Figs. 5.6 and 5.7. Figure 5.6 shows that the effect of the exoskeleton suit tracking the reference orientation is very good. And Fig. 5.7 shows the tracking error curve, from which we can see that the error is small, at 10^{-3} orders of magnitude.

In the same way, in order to analyze the controller's performance when the load changes, as shown in Sect. 5.4.1.1, it simulates the system in five cases and compares its position tracking error with its angle tracking error, the result being shown in Fig. 5.8. Contrasting the simulation curve of this section with that of the previous section, we can see the following: First, the oscillation of the position

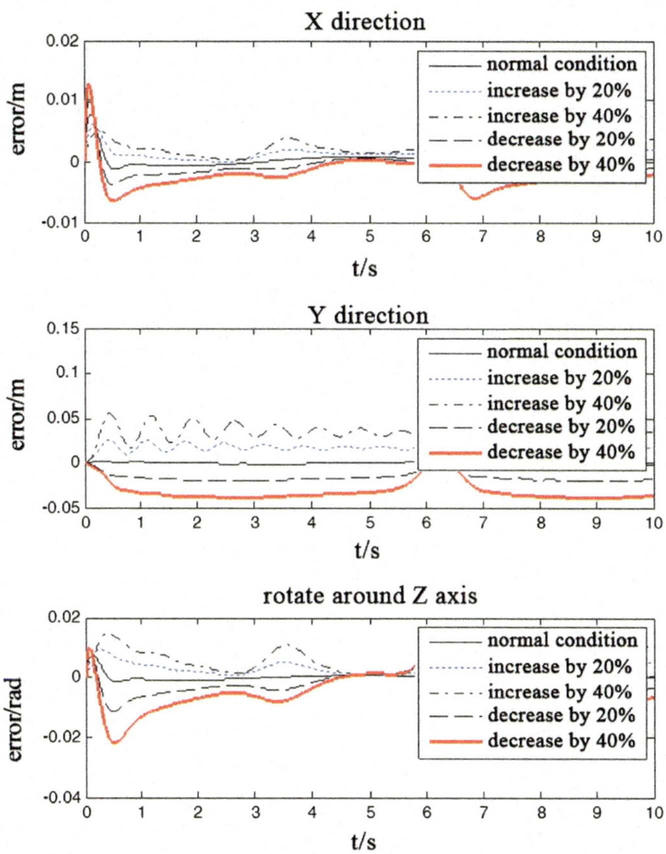

Fig. 5.8 Orientation tracking error contrast curve under the condition of load change

control's tracking curve based on the dynamic model is small, and its dynamic control effect is superior to position controller based on the static model; second, the steady-state error of the position controller based on the dynamic model is small; third, the structure of the position controller based on the dynamic model is complex, requiring precise mathematical model of system; and fourth, when the model error is big, the steady-state error increases, so the robustness of the position controller based on the dynamic model for the load variation is poor.

5.4.2 Simulation of Force Control Based on the Inner Position Loop

As shown in the last section, the inner position loop control can ensure the position error in a low level; therefore, we can add the force control outer loop to verify the

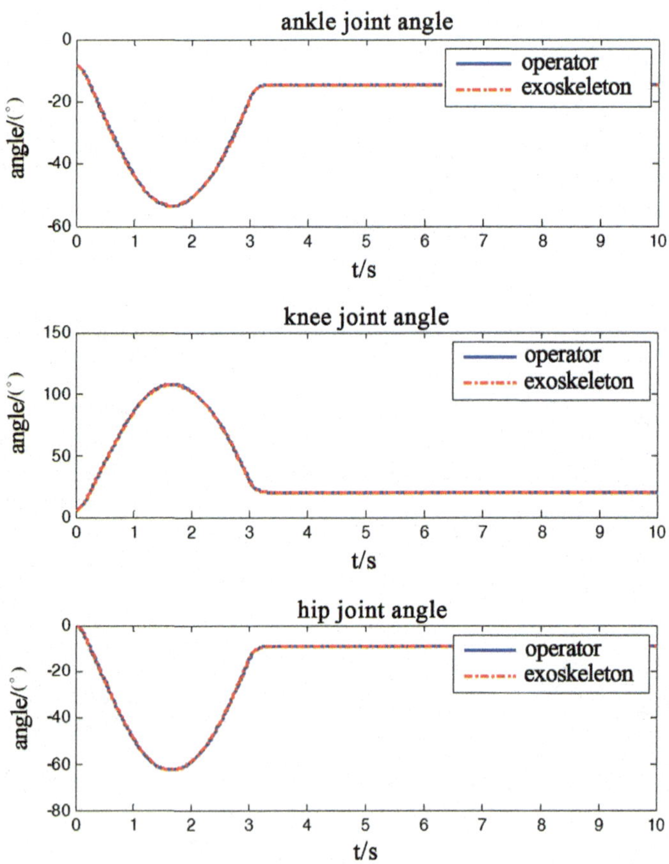

Fig. 5.9 Angle tracking curve of joint space

effect of force control. Here, we design the force controller using the position inner loop based on the static model, and the controller is as shown in Eq. (5.17).

In practical work of the exoskeleton suit, the input of the controller is the trunk's orientation of human, and its coupling with the exoskeleton suit will result in a human-machine interaction acting force $\boldsymbol{f}$. When multi-axis force sensor perceives $\boldsymbol{f}$ going into the force controller, it will produce the reference orientation signal of the inner position loop to control the exoskeleton suit's movement; when the position error decreases, f also becomes small, eventually making the exoskeleton suit move to the desired orientation and maintain at zero. In this process, the controller does not know the operator's desired orientation; it can only reflect the desired orientation through f indirectly, and the dynamic of the desired orientation and f is unknown.

However, during the simulation, the computer must give the operator's desired orientation and give the function relationship between specific desired orientation

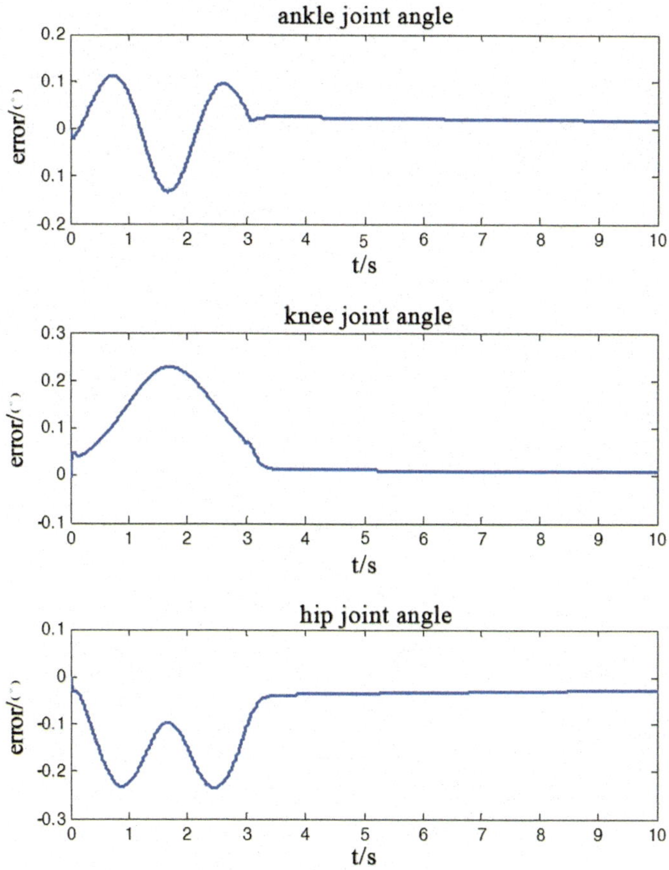

Fig. 5.10 Angle tracking error curve of joint space

and f, so the whole simulation process can be completed. If we use SimMechanics toolbox of the MATLAB to simulate, in the SimMechanics environment, first mechanical models of the operator (also considered as a multi-rigid-body) and the exoskeleton suit are established; the specific method is same as that in Sect. 4.4.

In the simulation, it uses the spring damping model as shown in Eq. (4.8) to describe the relationship between $\boldsymbol{f}$ and human-machine error, in which the spring coefficient $\boldsymbol{K}_{\mathrm{Pf}}$ and damping coefficient $\boldsymbol{K}_{\mathrm{Df}}$ are set in JSD. And choose:

$$\boldsymbol{K}_{\mathrm{Pf}} = \mathrm{diag}[5000 \quad 10000 \quad 250] \tag{5.34}$$

$$\boldsymbol{K}_{\mathrm{Df}} = \mathrm{diag}[100 \quad 500 \quad 5] \tag{5.35}$$

It is not easy to give desired orientation directly in the operator's mechanical model established with SimMechanics; on the contrary, it is easy to give the desired

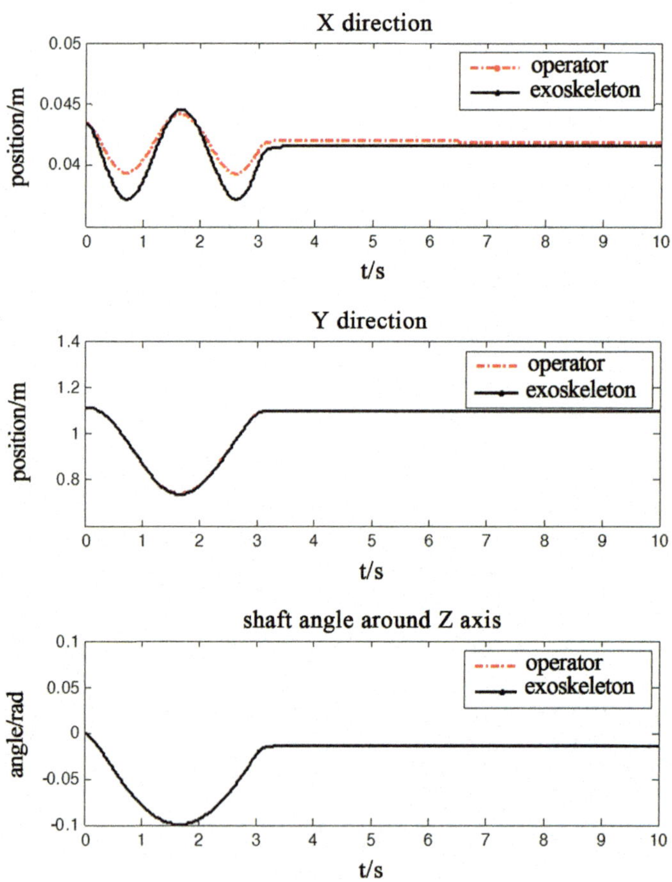

Fig. 5.11 Orientation tracking curve of operation space

trajectory of joint space, and the orientational transformation from the joint space to the operation space is generated by the internal mechanism of the SimMechanics, without needing the user's participation; therefore, what is given directly is the operator's reference trajectory in the joint space. In this simulation, two reference trajectories are given. Within 0–3 s, the joint angle moves following the trajectory shown from Eqs. (4.23) to (4.25). After 3 s, it will stabilize in the position of 3 s, and in order to make the reference trajectory realize smooth transition at 3 s, after generating the reference trajectory described above, through filtering with a first-order inertia term, we obtain the smooth reference trajectory and we set the time constant of the inertial term as 0.1. On the whole, this reference trajectory has simulated the process in which the operator drives the exoskeleton suit to do squatting and rising actions and then stands still for a period of time.

The position controller based on the static model is shown in Eqs. (5.2) and (5.3), and the force controller is shown in Eq. (5.20). During simulation, the

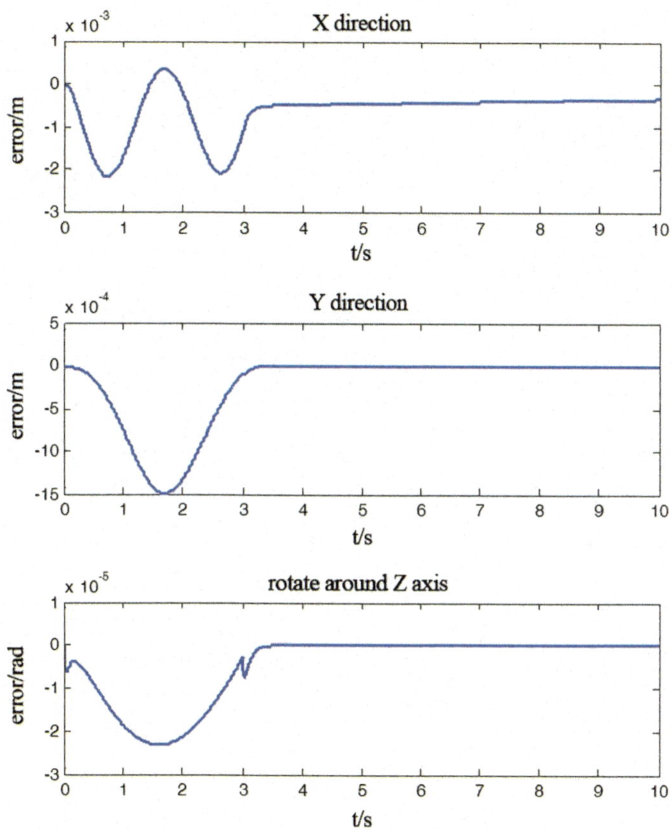

Fig. 5.12 Orientation tracking error curve of operation space

controller parameters of the inner position loop are selected as shown in Eqs. (5.30) and (5.31), and the parameters of the force controller are as follows:

$$K_{\mathrm{P}} = \mathrm{diag}[\,10 \quad 500 \quad 80\,] \tag{5.36}$$

$$K_{\mathrm{I}} = \mathrm{diag}[\,1 \quad 10 \quad 6.5\,] \tag{5.37}$$

The simulation results are shown in Figs. 5.9 and 5.18. Figure 5.9 shows the joint angle tracking curve, from the top to the bottom being the ankle, knee, and hip joints, respectively, namely q_1, q_2, q_3, in which the solid line represents the operator's desired joint angle in joint space and the dot-dash line represents the exoskeleton suit's joint angle in joint space. From the figure, we can see that the two curves almost overlap. Figure 5.10 shows the tracking error curve. The two figures illustrate that the tracking effect of the exoskeleton suit on the human body movement is good, and the tracking error is small; if we use the degree as the unit, the error is at 10^{-1} orders of magnitude.

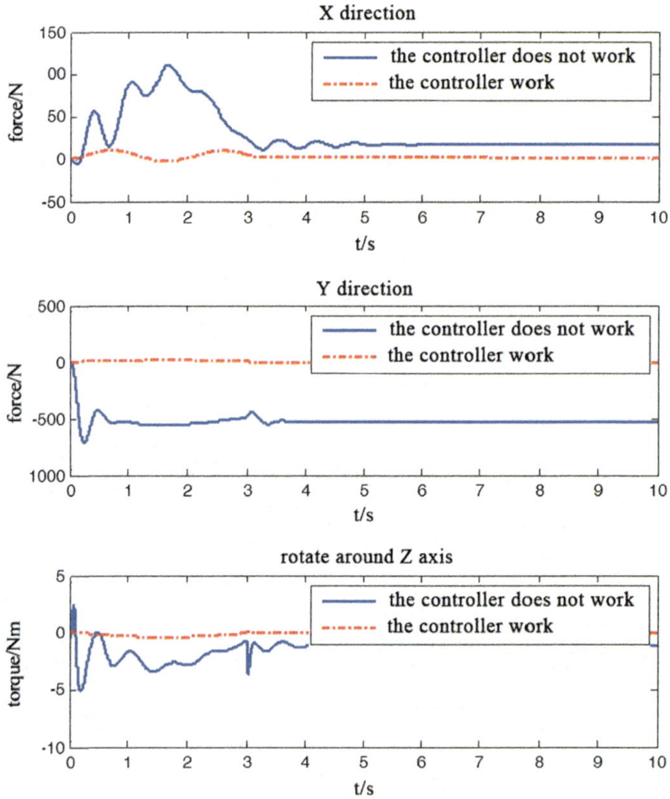

Fig. 5.13 Human-machine acting force of operation space

Figure 5.11 shows the operator's position and the exoskeleton suit's orientation curve in the operation space, from top to bottom being the X direction position, Y direction position, and the rotation angle around the Z-axis, respectively, namely p_x, p_y, θ. Its results are converted from the data into Fig. 5.9 according to the geometrical relationship between joint space and operation space, in which the solid line represents the position of the trunk's center of gravity of the operator in the operation space, namely $\boldsymbol{x}_h$, and the dot-dash line represents the position of the trunk's center of gravity of the exoskeleton suit in the operation space, namely $\boldsymbol{x}_e$. From the figure, we can see that to estimate the orientation of the operator according to the human-machine acting force, $\boldsymbol{f}$ is relatively accurate, and it realizes the exoskeleton suit tracking the human body movement in the operation space successfully. In Fig. 5.11, the deviation in X direction looks bigger, but in fact because the movement of the operator and exoskeleton suit in the X direction is very small, the actual deviation between them remains at 10^{-3} orders of magnitude, and the specific deviation is shown in Fig. 5.12.

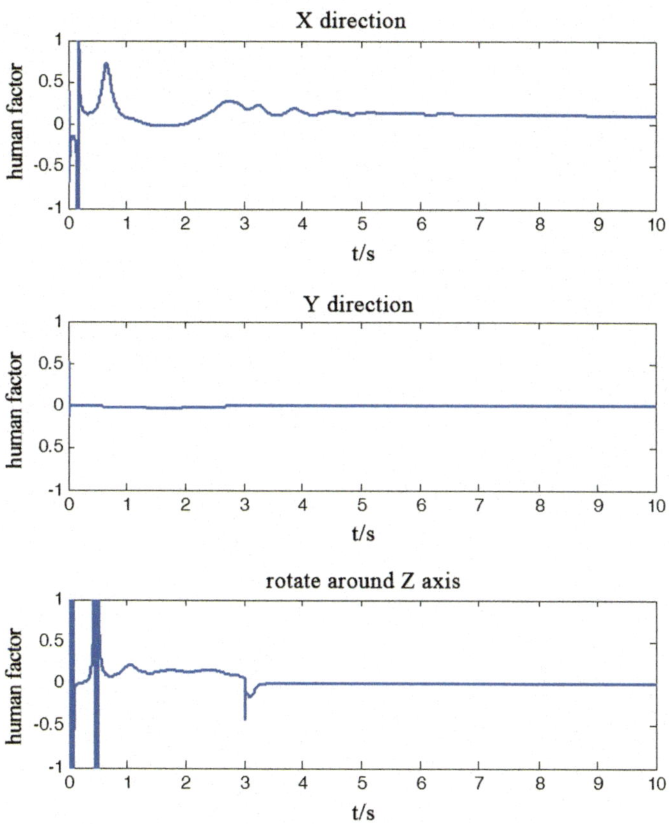

Fig. 5.14 Human factor of operation space

Figure 5.13 shows the human-machine acting force curve in operation space, from top to bottom being the human-machine force in X direction, the human-machine force in Y direction, and the human-machine torque around the Z-axis, respectively, namely f_x,f_y, T_z, and it shows the human-machine acting force when the actuator is at work and it does not work, respectively. That the actuator does not work can also be understood as not adding force and inner position loop control, when $\boldsymbol{T}_a$ is zero, and the exoskeleton suit moves completely under the function of human-machine acting force $\boldsymbol{f}$. In the figure, the solid line represents the $\boldsymbol{f}$ at this time and the dot-dash line represents the human-machine force $\boldsymbol{f}$ when the controller is working. Namely, the control torque of the exoskeleton suit includes not only the human-machine force $\boldsymbol{f}$ exerted by the operator but also the control force torque $\boldsymbol{T}_a$ exerted by the controller. Figure 5.13 shows that when the controller is working, the human-machine acting force/torque is greatly reduced, which means that the energy consumption of the operator is greatly reduced. In the figure, what is shown before 3 s is the human-machine acting force in squatting and

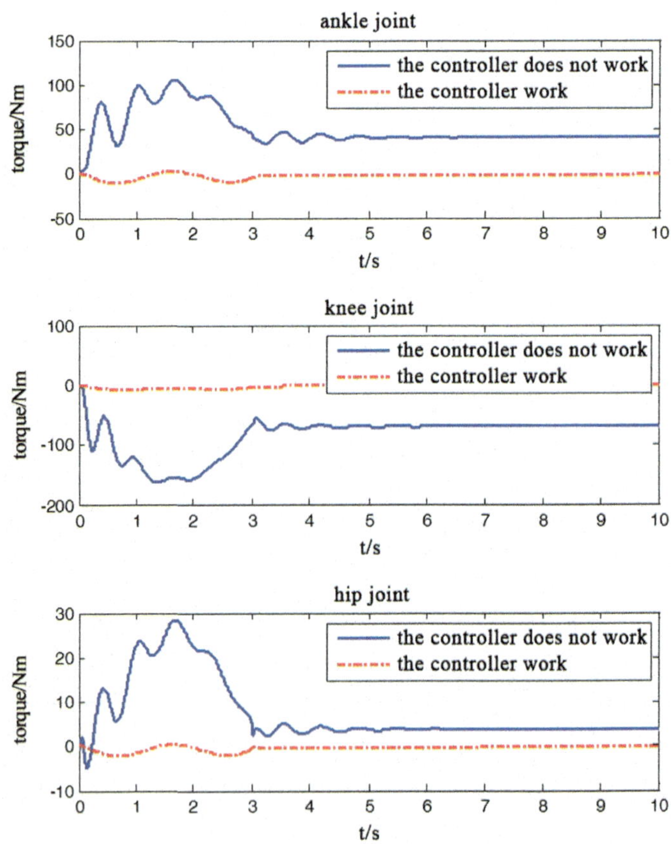

Fig. 5.15 Human-machine acting force of joint space

rising action, and after 3 s, it is the human-machine acting force in standing still phase after the squatting and rising action, from which we can see that in both dynamic and static processes, the designed controller can significantly reduce the human-machine acting force, and under the static condition, human-machine acting force converges to zero, illustrating that in static condition, the exoskeleton suit under the control of the controller can stand still itself, without the need of operator's extra force. Figure 5.14 shows the corresponding human factor $\boldsymbol{HF}$, most of which is within ±0.3, and in the later period, it even approaches zero, corresponding to the maximum human factor $\boldsymbol{HFM} = [0.0977 \quad 0.0207 \quad 0.1012]$, also very small, illustrating that the human force accounts for very small proportion in movement of the exoskeleton suit. Figures 5.15 and 5.16 show the human-machine acting force contrast curve and human factor curve in the joint space, respectively, corresponding to the maximum human factor $\boldsymbol{HFM} = [0.1000 \quad 0.0450 \quad 0.0768]$, illustrating the same problem.

To illustrate the performance of the designed controller, we specially simulate the human-machine system without adding force and inner position loop controller

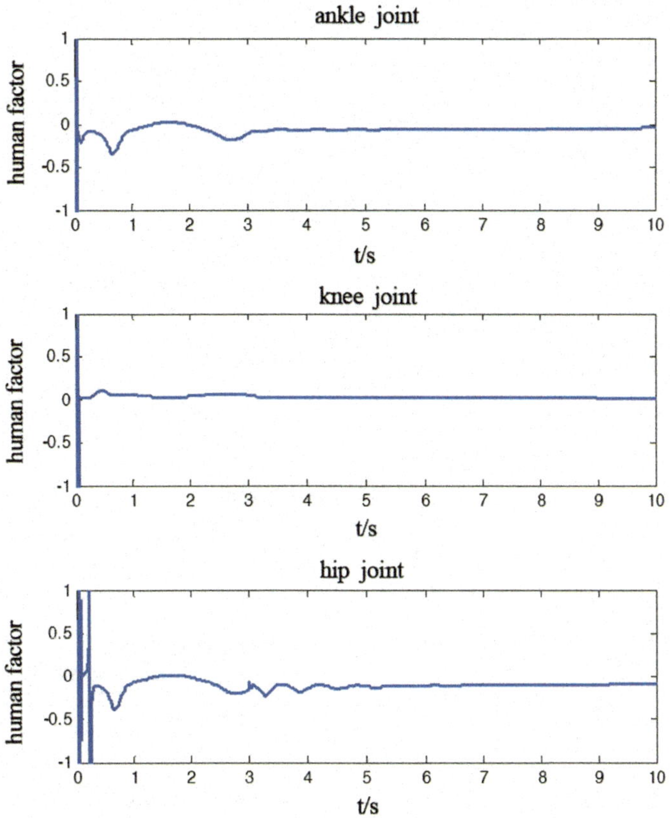

Fig. 5.16 Human factor of joint space

(or when the controller does not work), during the tracking of the exoskeleton suit for the human body's trajectory as shown in Fig. 5.17 and the tracking error as shown in Fig. 5.18, from which we can see that the tracking error of the three joints that have larger tracking error relatively reaches 13°, 26°, and 14°, respectively. From Figs. 5.9, 5.10, 5.11, 5.12, 5.13, 5.14, 5.15, 5.16, 5.17, and 5.18 it is shown that the designed controller has reached the desired control purpose, and it is effective and feasible.

In this section, the simulation is only for the support leg. For the control of the swing leg, this control mode can also be used, but the installation position of the sensor needs rethinking; for example, if it is installed on the ankle, the corresponding Jacobian matrix needs to change accordingly. But we can also use other control modes and even do not exert any control, which is completely controlled by

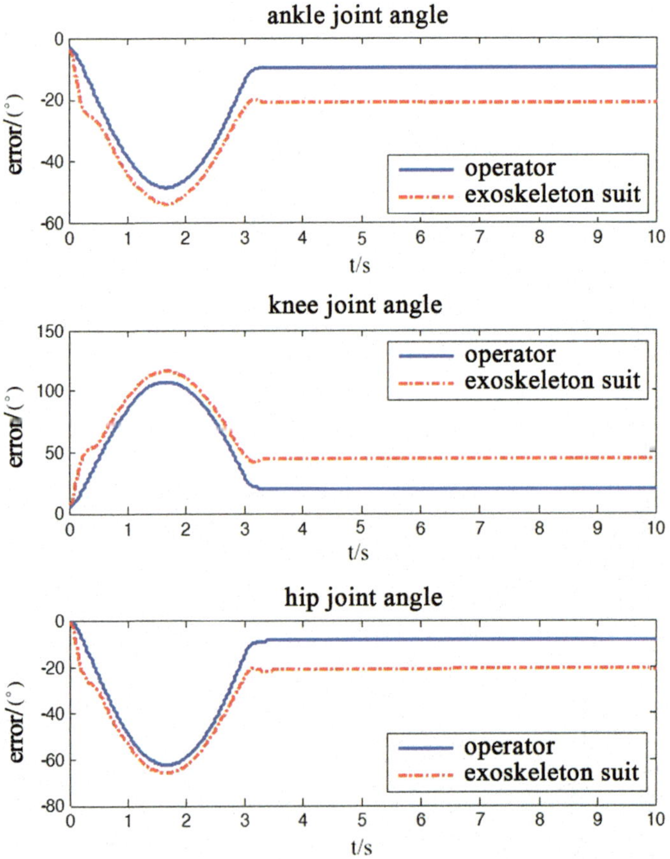

Fig. 5.17 Joint angle tracking of the exoskeleton suit for the human body without adding the controller

the operator. This is because the control goal of the exoskeleton suit is to reduce the energy consumption of the human body when it bears load, and the consumption is mainly from the support leg.

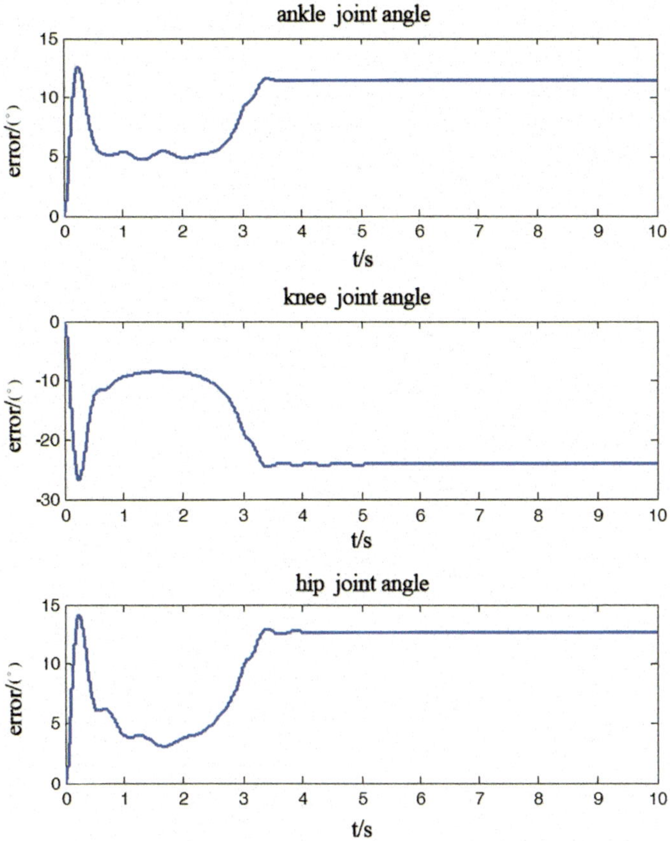

Fig. 5.18 Joint angle tracking error of the exoskeleton suit for the human body without adding the controller

Table 5.1 Human-machine interaction model parameters

Parameters	Parameter1	Parameter2	Parameter3	Parameter4
K_{pfx}	5000	2500	1250	1000
K_{dfx}	100	50	25	20
K_{pfy}	10000	5000	2500	2000
K_{dfy}	500	250	125	100
K_{pfrz}	250	125	62.5	50
K_{dfrz}	5	2.5	1.25	1

5.4.2.1 The Influence of Human-Machine Interaction Model Parameters on Human-Machine Force

Parameters of the human-machine interaction model shown in Eq. (4.8) represent the interaction between human and the exoskeleton suit. For different operators, such as the strong and weak people, these parameters are different. Moreover, from Sect. 2.5, we can see that the order of human-machine interaction model can also be the first order, or the third order, and the structure can even be nonlinear. This section does not consider other structures temporarily, but it studies the simulation results under the condition of human-machine interaction model shown in Eq. (4.8) and the parameter changing.

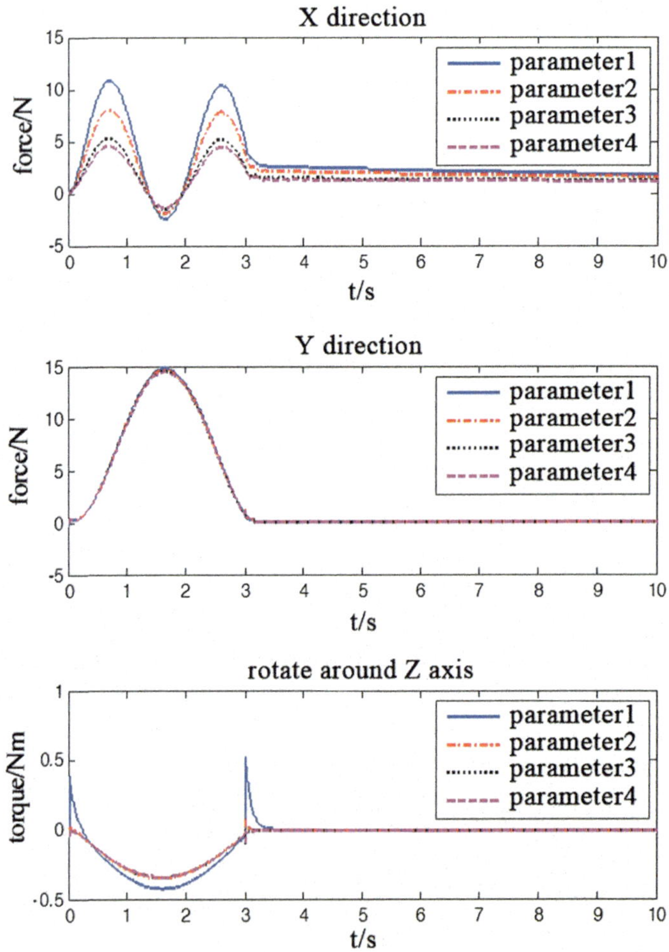

Fig. 5.19 Human-machine acting force of operation space when human-machine interaction model parameters change

Please refer to Eqs. (4.26) and (4.27) for the definition of human-machine interaction model parameters, and we choose four different cases to simulate, as shown in Table 5.1.

When human-machine interaction model parameters change, the trajectory tracking conditions in joint space and operation space is good, and the tracking

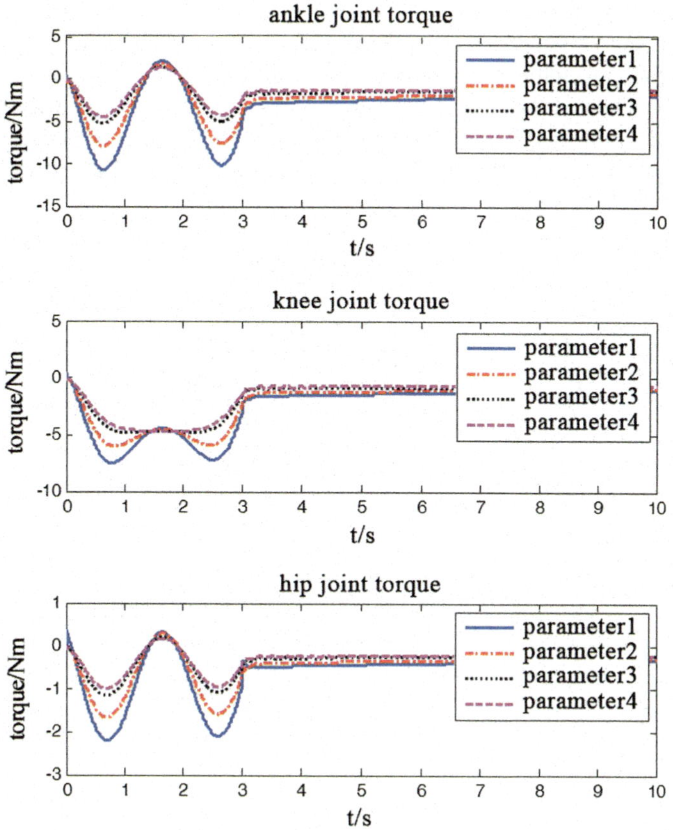

Fig. 5.20 Human-machine acting force of joint space when human-machine interaction model parameters change

Table 5.2 Maximum human factor when human-machine interaction model parameters change

Parameters	X direction	Y direction	Rotating around the Z-axis	Ankle joint	Knee joint	Hip joint
Parameter1	0.0977	0.0207	0.1012	0.1005	0.0462	0.0780
Parameter2	0.0724	0.0205	0.0681	0.0745	0.0370	0.0583
Parameter3	0.0477	0.0203	0.0671	0.0491	0.0295	0.0401
Parameter4	0.0407	0.0202	0.0670	0.0419	0.0287	0.0350

error is relatively small, similar to Fig. 5.9; therefore, we no longer show one by one. Here, we only show the changing curve of human-machine acting force when human-machine interaction model parameters change, as shown in Figs. 5.19 and 5.20. Parameters in Figs. 1, 2, 3, and 4, respectively, correspond to the simulation results under the conditions of parameters in Tables 1, 2, 3, and 4. Figure 5.19 shows the human-machine acting force curve in operation space when human-machine interaction model parameters change, from which we can see that in the X direction, with the gradual decrease in human-machine interaction model parameter, the human-machine acting force also gradually becomes smaller. The reason is that position tracking error decreases gradually in the X direction, while in the Y direction and the direction of rotating Z-axis, the position tracking error is small all the time, so the changes of human-machine force is not large either, and the corresponding human factor ***HF*** curve is similar to Fig. 5.14, so we no longer

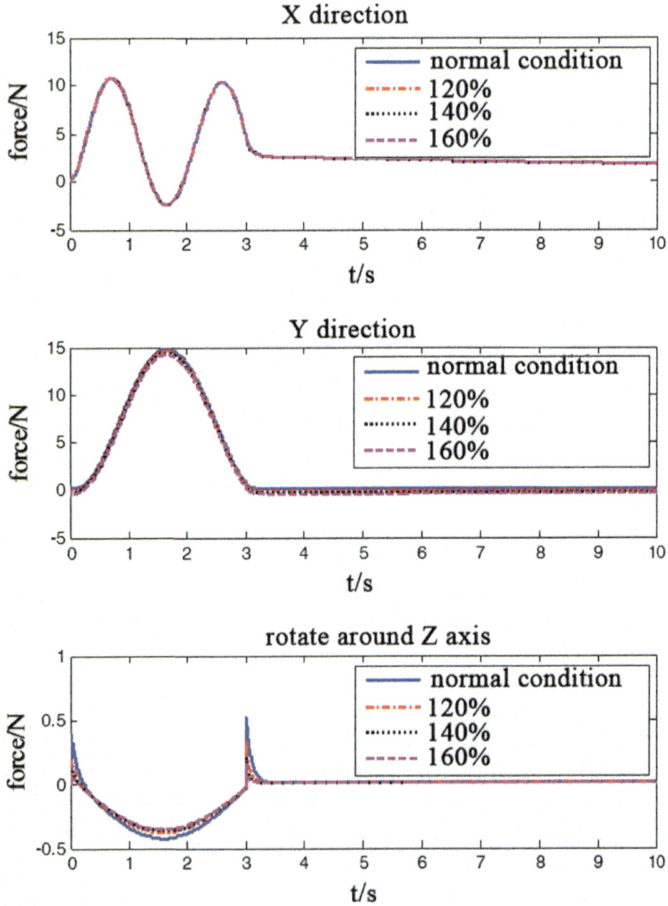

Fig. 5.21 Human-machine acting force of operation space when load changes

give. Figure 5.20 shows the human-machine acting force curve of joint space. Because the human-machine acting force of joint space is converted into human-machine acting force of operation space, contrasted with Fig. 5.19, we can see that the relative large change of operation space in certain direction can cause changes of joint space at the same time in three directions. The changes of human factor ***HF*** of joint space are also correspondingly consistent, similar to Fig. 5.16, so we no longer give here.

Table 5.2 gives the maximum human factor ***HFM*** when human-machine interaction model parameters change, from which we can see that when the human-machine interaction model parameters change, both in operation space and in joint space, the maximum human factor is very small and the variation is also very small, less than 0.06. These simulation results illustrate from different aspects that the robustness of the force controller based on the inner position loop is relatively strong for changes of the human-machine interaction model parameters, or changes of the environment (the operator).

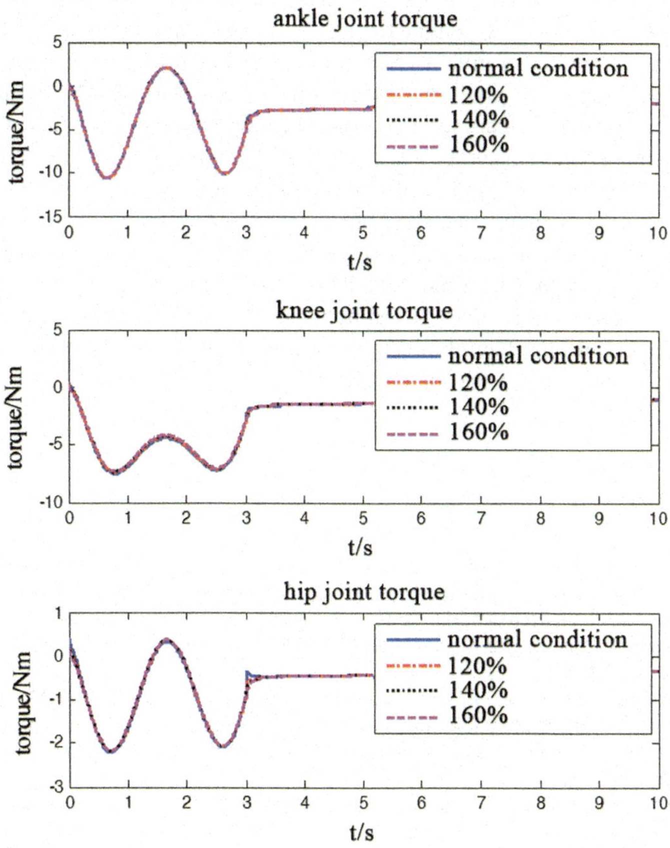

Fig. 5.22 Human-machine acting force of joint space when load changes

Table 5.3 Maximum human factor when the load changes

Load	X direction	Y direction	Rotating around the Z-axis	Ankle joint	Knee joint	Hip joint
Normal conditions	0.0977	0.0207	0.1012	0.1005	0.0462	0.0780
120%	0.0975	0.0204	0.0739	0.1002	0.0457	0.0772
140%	0.0974	0.0201	0.0712	0.1001	0.0453	0.0769
160%	0.0973	0.0198	0.0699	0.1000	0.0450	0.0768

5.4.2.2 The Influence of Load Changes on Human-Machine Force

Similar to Sect. 4.4.2, the effect of the controller is tested by studying the changes of human-machine force when the load changes. The law of load changes is the same as that in Sect. 4.4.2. The simulation results are shown in Figs. 5.21 to 5.22. What is shown in Fig. 5.21 is the human-machine acting force changing curve in operation space when the load changes, from which we can see that although the load increase, the human-machine force changes little. The human factor ***HF*** in operation space when the load changes is similar to Fig. 5.14, and the human factor has hardly changed, so we no longer give. In joint space, there are the same changes, as shown in Fig. 5.22. Table 5.3 gives the maximum human factor ***HFM*** when the load varies, from which we can see that when the load changes, the variation in the maximum human factor is very small, with the variation at 10^{-4} orders of magnitude.

From the perceptual intuition, if without function of the exoskeleton suit, when the load borne by human increases, human needs to exert more force on the load to offset the gravity of the load. However, through the simulation, we find that after using the exoskeleton suit, when the load increases, the acting force exerted by the operator is similar to the force exerted by the operator without load, illustrating that the designed controller can eliminate the impact of load changing, and namely, it has strong robustness for the load change.

References

1. Zhang Guangli. Force Control System and Advanced Control Strategy Research for Open Structure Robot [D]. Shanghai Jiaotong University, Doctoral Dissertation, Shanghai: 2003.
2. Volpe, Richard, Pradeep Khosla. A Theoretical and Experimental Investigation of Explicit Force Control Strategies for Manipulators [J]. IEEE Transactions on Automatic Control, 1993, 38 (11): 1634–1650.
3. Yin Yuehong, Wei Zhongxin, Huang Xiaoxi. Force Sensing and Force Control Technology for Intelligent Machine System [M]. Beijing: National Defence Industry Press, 2002.
4. Siciliano, Bruno, Luigi Villani. Robot Force Control [M]. Boston/Dordrecht/London: Kluwer Academic Publishers, 1999.
5. Winter, D A. Biomechanics of Human Movement [M]. New York: John Wiley and Sons, 1979.

Chapter 6
Impedance Control of Exoskeleton Suit

6.1 Introduction

Impedance control is also one of the classical force control methods. It is according to the relationship between position of the machine end (or velocity) and the acting force of the end part, and by adjusting feedback position error, velocity error or stiffness to achieve the purpose of control force [1]. The impedance control method is first put forward by Hogon in 1985, and after research by many scholars, a variety of control methods based on impedance control have been developed and applied to various controlled objects [2–5]. Impedance control is done by adjusting the dynamic relation between the location of the end executor and the contact force to realize the compliance control. This method provides a unified method for collision avoidance and constrained and unconstrained movement. Its advantages include it requires little offline mission planning, it has good robustness for the disturbance and uncertainty, and it can realize the stable transformation of the system from the unconstrained movement to constrained movement, so the impedance control is considered more suitable for completing the assembly work. When the impedance control is applied to the control of the exoskeleton suit, it can adjust the dynamic relationship between the terminal of the exoskeleton suit and human-machine acting force by adjusting the parameter of the impedance control, whose physical meaning is to adjust people's comfort level, but because there are no corresponding experiments currently, we cannot make an accurate definition for the human body comfort level; therefore, in the book, we only study the possibility of its implementation, and for the comfort level, we will not study, neither in other chapters.

However, in the impedance control, the reference trajectory is usually known; when it contacts with the environment, we need to use force information to estimate the reference trajectory correction [1, 4, 6]. The impedance control of the exoskeleton suit is different from that of the common robot, whose reference trajectory is unknown, so we need to estimate the reference trajectory according to the force information, not just the reference trajectory correction. At the same time, for

Z. Yang et al., *Force Control Theory and Method of Human Load Carrying Exoskeleton Suit*, DOI 10.1007/978-3-662-54144-9_6

the impedance control method of the exoskeleton suit introduced in this chapter, the exoskeleton suit's reference speed and reference acceleration also needs estimating. If we only estimate the reference trajectory, and through the finite difference method to obtain the signals of reference speed and the reference acceleration, the noise signal contained in the force feedback information can lead to that the reference speed and the reference acceleration signals are unavailable. For the above problem, this book introduces an impedance control method based on tracking differentiator. When we use this method, first design the impedance controller of the exoskeleton suit, set the desired human-machine acting force to be zero, then use multi-dimensional force/torque sensor to measure the force/torque information of human acting on the exoskeleton suit in operation space, construct the reference trajectory signal of the impedance controller, and make use of the tracking differentiator to get signals of the smooth reference speed and reference acceleration. Ultimately, control the exoskeleton suit to follow the human body's movement, and make human-machine acting force to be zero, reducing the energy consumption of human body. This method solves the problem of estimating reference speed and reference acceleration. This book gives simulation results, illustrates the feasibility and effectiveness of the designed controller, and analyzes the robustness of the controller under the conditions of the estimation of body stiffness parameter not being accurate and model parameter being uncertain.

6.2 Impedance Control of Exoskeleton Suit

If we do not consider movement friction and refer to Eq. (2.74), then the kinetic model of the exoskeleton suit system can be expressed as [7]

$$\boldsymbol{H}(\boldsymbol{q})\ddot{\boldsymbol{q}} + \boldsymbol{C}(\boldsymbol{q},\dot{\boldsymbol{q}})\dot{\boldsymbol{q}} + \boldsymbol{G}(\boldsymbol{q}) = \boldsymbol{T}_{\mathrm{a}} + \boldsymbol{T}_{\mathrm{hm}} \tag{6.1}$$

The relationship between human-machine torque $\boldsymbol{T}_{\mathrm{hm}}$ of joint space and the generalized force $\boldsymbol{f}$ of operation space is

$$\boldsymbol{T}_{\mathrm{hm}} = \boldsymbol{J}^{\mathrm{T}}(\boldsymbol{q})\boldsymbol{f} \tag{6.2}$$

Let $\boldsymbol{x}$, $\dot{\boldsymbol{x}}$, and $\ddot{\boldsymbol{x}}$ represent the orientation vector, velocity vector, and acceleration vector (a case of the generalized coordinate vector) of the center of gravity of the exoskeleton suit's trunk in operation space, respectively, then there are

$$\dot{\boldsymbol{x}} = \boldsymbol{J}(\boldsymbol{q})\dot{\boldsymbol{q}} \tag{6.3}$$

$$\ddot{\boldsymbol{x}} = \boldsymbol{J}(\boldsymbol{q})\ddot{\boldsymbol{q}} + \dot{\boldsymbol{J}}(\boldsymbol{q})\dot{\boldsymbol{q}} \tag{6.4}$$

After substituting Eqs. (6.2), (6.3), and (6.4) into Eq. (6.1), then, we obtain the dynamic equation of the exoskeleton suit in operation space

$$\boldsymbol{A}(\boldsymbol{q})\ddot{\boldsymbol{x}} + \boldsymbol{B}(\boldsymbol{q}, \dot{\boldsymbol{q}})\dot{\boldsymbol{x}} + \boldsymbol{D}(\boldsymbol{q}) = \boldsymbol{u} + \boldsymbol{f} \tag{6.5}$$

In the equation,

$$\boldsymbol{A}(\boldsymbol{q}) = \boldsymbol{J}^{-\mathrm{T}}(\boldsymbol{q})\boldsymbol{H}(\boldsymbol{q})\boldsymbol{J}^{-1}(\boldsymbol{q}) \tag{6.6}$$

$$\boldsymbol{B}(\boldsymbol{q}) = \boldsymbol{J}^{-\mathrm{T}}(\boldsymbol{q})\boldsymbol{C}(\boldsymbol{q}, \dot{\boldsymbol{q}})\boldsymbol{J}^{-1}(\boldsymbol{q}) - \boldsymbol{A}(\boldsymbol{q})\dot{\boldsymbol{J}}(\boldsymbol{q})\boldsymbol{J}^{-1}(\boldsymbol{q}) \tag{6.7}$$

$$\boldsymbol{D}(\boldsymbol{q}) = \boldsymbol{J}^{-\mathrm{T}}(\boldsymbol{q})\boldsymbol{G}(\boldsymbol{q}) \tag{6.8}$$

$$\boldsymbol{u} = \boldsymbol{J}^{-\mathrm{T}}(\boldsymbol{q})\boldsymbol{T}_{\mathrm{a}} \tag{6.9}$$

Design the controller based on the inverse dynamic method

$$\boldsymbol{T}_{\mathrm{a}} = \hat{\boldsymbol{H}}(\boldsymbol{q})\boldsymbol{\alpha} + \hat{\boldsymbol{C}}(\boldsymbol{q}, \dot{\boldsymbol{q}})\dot{\boldsymbol{q}} + \hat{\boldsymbol{G}}(\boldsymbol{q}) - \boldsymbol{J}^{\mathrm{T}}(\boldsymbol{q})\boldsymbol{f} \tag{6.10}$$

$$\boldsymbol{\alpha} = \boldsymbol{J}^{-1}(\boldsymbol{q})(\boldsymbol{\tau} - \dot{\boldsymbol{J}}(\boldsymbol{q}, \dot{\boldsymbol{q}})\dot{\boldsymbol{q}}) \tag{6.11}$$

$$\boldsymbol{\tau} = \ddot{\boldsymbol{x}}_{\mathrm{c}} + \boldsymbol{K}_{\mathrm{Mp}}{}^{-1}(\boldsymbol{K}_{\mathrm{Dp}}\Delta\dot{\boldsymbol{x}} + \boldsymbol{K}_{\mathrm{Pp}}\Delta\boldsymbol{x} + \boldsymbol{f}) \tag{6.12}$$

In the equations, $\hat{\boldsymbol{H}}(\boldsymbol{q})$, $\hat{\boldsymbol{C}}(\boldsymbol{q}, \dot{\boldsymbol{q}})$, and $\hat{\boldsymbol{G}}(\boldsymbol{q})$ are the estimated value of $\boldsymbol{H}(\boldsymbol{q})$, $\boldsymbol{C}(\boldsymbol{q}, \dot{\boldsymbol{q}})$, and $\boldsymbol{G}(\boldsymbol{q})$, respectively; $\boldsymbol{x}_{\mathrm{c}}$, $\dot{\boldsymbol{x}}_{\mathrm{c}}$, and $\ddot{\boldsymbol{x}}_{\mathrm{c}}$ are the orientation vector, velocity vector, and acceleration vector of the reference trajectory, and $\Delta\boldsymbol{x} = \boldsymbol{x}_{\mathrm{c}} - \boldsymbol{x}$, $\Delta\dot{\boldsymbol{x}} = \dot{\boldsymbol{x}}_{\mathrm{c}} - \dot{\boldsymbol{x}}$, and $\Delta\ddot{\boldsymbol{x}} = \ddot{\boldsymbol{x}}_{\mathrm{c}} - \ddot{\boldsymbol{x}}$; $\boldsymbol{K}_{\mathrm{Mp}}$, $\boldsymbol{K}_{\mathrm{Dp}}$, and $\boldsymbol{K}_{\mathrm{Pp}}$ are the ideal positive definite inertia coefficient matrix, damping coefficient matrix, and rigidity coefficient matrix, respectively.

After substituting the equations from (6.9) to (6.12) into Eq. (6.5), then, we obtain the dynamic equation of the closed-loop system

$$\Delta\ddot{\boldsymbol{x}} + \boldsymbol{K}_{\mathrm{Mp}}{}^{-1}(\boldsymbol{K}_{\mathrm{Dp}}\Delta\dot{\boldsymbol{x}} + \boldsymbol{K}_{\mathrm{Pp}}\Delta\boldsymbol{x} + \boldsymbol{f}) = \hat{\boldsymbol{A}}(\boldsymbol{q})^{-1}(\Delta\boldsymbol{A}(\boldsymbol{q})\ddot{\boldsymbol{x}} + \Delta\boldsymbol{B}(\boldsymbol{q}, \dot{\boldsymbol{q}})\dot{\boldsymbol{x}} + \Delta\boldsymbol{D}(\boldsymbol{q})) \tag{6.13}$$

In the equations, $\Delta\boldsymbol{A}(\boldsymbol{q}) = \boldsymbol{A}(\boldsymbol{q}) - \hat{\boldsymbol{A}}(\boldsymbol{q})$, $\Delta\boldsymbol{B}(\boldsymbol{q}) = \boldsymbol{B}(\boldsymbol{q}) - \hat{\boldsymbol{B}}(\boldsymbol{q})$, and $\Delta\boldsymbol{D}(\boldsymbol{q}) = \boldsymbol{D}(\boldsymbol{q}) - \hat{\boldsymbol{D}}(\boldsymbol{q})$, $\hat{\boldsymbol{A}}(\boldsymbol{q})$, $\hat{\boldsymbol{B}}(\boldsymbol{q})$, and $\hat{\boldsymbol{D}}(\boldsymbol{q})$ are the estimated value of $\boldsymbol{A}(\boldsymbol{q})$, $\boldsymbol{B}(\boldsymbol{q})$, and $\boldsymbol{D}(\boldsymbol{q})$, respectively, and

$$\hat{\boldsymbol{A}}(\boldsymbol{q}) = \boldsymbol{J}^{-\mathrm{T}}(\boldsymbol{q})\hat{\boldsymbol{H}}(\boldsymbol{q})\boldsymbol{J}^{-1}(\boldsymbol{q}) \tag{6.14}$$

$$\hat{\boldsymbol{B}}(\boldsymbol{q}) = \boldsymbol{J}^{-\mathrm{T}}(\boldsymbol{q})\hat{\boldsymbol{C}}(\boldsymbol{q}, \dot{\boldsymbol{q}})\boldsymbol{J}^{-1}(\boldsymbol{q}) - \hat{\boldsymbol{A}}(\boldsymbol{q})\dot{\boldsymbol{J}}(\boldsymbol{q})\boldsymbol{J}^{-1}(\boldsymbol{q}) \tag{6.15}$$

$$\hat{\boldsymbol{D}}(\boldsymbol{q}) = \boldsymbol{J}^{-\mathrm{T}}(\boldsymbol{q})\hat{\boldsymbol{G}}(\boldsymbol{q}) \tag{6.16}$$

In ideal conditions, $\Delta\boldsymbol{A}(\boldsymbol{q}) = \Delta\boldsymbol{B}(\boldsymbol{q}) = \Delta\boldsymbol{D}(\boldsymbol{q}) = 0$. Then, the behavior of the human-machine acting force and the exoskeleton suit meets the target impedance relationship:

$$-\boldsymbol{f} = \boldsymbol{K}_{\mathrm{Mp}}\Delta\ddot{\boldsymbol{x}} + \boldsymbol{K}_{\mathrm{Dp}}\Delta\dot{\boldsymbol{x}} + \boldsymbol{K}_{\mathrm{Pp}}\Delta\boldsymbol{x} \tag{6.17}$$

Through proper design, $\boldsymbol{K}_{\mathrm{Mp}}$, $\boldsymbol{K}_{\mathrm{Dp}}$, and $\boldsymbol{K}_{\mathrm{Pp}}$ can make human swing the exoskeleton suit easily, achieving compliance control or achieving the compliance control of meeting the target impedance.

After observing carefully, let $\boldsymbol{K}_{\mathrm{Pp}} = 0$, and in the ideal steady state, $\boldsymbol{f} = 0$.

The reference trajectory of the exoskeleton suit is actually the movement trajectory of the operator; however, to measure the operator's movement trajectory actually is very difficult, which needs the complex sensors to be installed on the human body, but the measuring accuracy of current technology is not high, and installing sensors on the human body will reduce people's comfort degree, and make the wearing of exoskeleton suit inconvenient. As stated earlier, the multi-dimensional force/torque sensor adopted by our research is through the back's straps of the human body to couple with the human body, which is convenient to use, and through the measurement of the sensors we can get the human-machine force $\boldsymbol{f}$ to estimate the reference trajectory.

As shown in Eq. (4.9), the simplest human-machine acting force can be modeled as [6]

$$\boldsymbol{f} = \boldsymbol{K}_{\mathrm{Pf}}(\boldsymbol{x}_{\mathrm{c}} - \boldsymbol{x}) \tag{6.18}$$

In the equation, $\boldsymbol{K}_{\mathrm{Pf}}$ represents the human body stiffness (stiffness coefficient matrix). So the reference trajectory can be estimated by Eq. (6.19)

$$\hat{\boldsymbol{x}}_{\mathrm{c}} = \frac{\boldsymbol{f}}{\hat{\boldsymbol{K}}_{\mathrm{Pf}}} + \boldsymbol{x} \tag{6.19}$$

In the equation, $\hat{\boldsymbol{K}}_{\mathrm{Pf}}$ represents the estimation of the human body stiffness. Although the estimation of $\hat{\boldsymbol{K}}_{\mathrm{Pf}}$ may not be accurate, as long as $\hat{\boldsymbol{K}}_{\mathrm{Pf}}$ meets certain conditions, the system's stability still can be guaranteed. After substituting Eq. (6.19) into Eq. (6.17), we can get

$$\boldsymbol{K}_{\mathrm{Mp}}\ddot{\boldsymbol{f}} + \boldsymbol{K}_{\mathrm{Dp}}\dot{\boldsymbol{f}} + (\boldsymbol{K}_{\mathrm{Pp}} + \hat{\boldsymbol{K}}_{\mathrm{Pf}})\boldsymbol{f} = 0 \tag{6.20}$$

Therefore, as long as $\boldsymbol{K}_{\mathrm{Dp}}$ and $\boldsymbol{K}_{\mathrm{Pp}} + \hat{\boldsymbol{K}}_{\mathrm{Pf}}$ are positive definite, $\boldsymbol{f}$ can be guaranteed converging to zero.

6.3 Impedance Control Based on Tracking Differentiator

However, by Eq. (6.12), we can see that the designed controller not only needs reference orientation signal (reference trajectory signal), but also needs signals of reference speed and reference acceleration. Ideally, the reference speed and reference acceleration can be obtained by the difference method. However, the measurement of $\boldsymbol{f}$ is bound to contain measurement noise, and the difference method will amplify the noise, making signals of reference speed and reference acceleration degrade, so it is not applicable.

Han Jingqing has constructed the mechanism of tracking discrete input signal and extracting the approximate differential signal making use of the second-order steepest switch system, has put forward the concept of nonlinear tracking differentiator, and has proved that the differential signal obtained by tracking differentiator is the smooth approximation of generalized derivative of the input signal [8, 9].

In this book, we use two-stage tracking differentiator to get smooth reference orientation, reference speed, and reference acceleration. Use the reference orientation $\hat{\boldsymbol{x}}_c$ generated by Eq. (6.9)'s estimation as input of the first stage of tracking differentiator, output the tracking reference orientation $\hat{\boldsymbol{x}}_c^1$ and reference speed signal $\hat{\dot{\boldsymbol{x}}}_c^1$, and then use $\hat{\dot{\boldsymbol{x}}}_c^1$ as input of the second stage of tracking differentiator; get the tracking reference speed signal $\hat{\dot{\boldsymbol{x}}}_c^2$ and reference acceleration signal $\hat{\ddot{\boldsymbol{x}}}_c^2$, and eventually get $\hat{\boldsymbol{x}}_c^1$, $\hat{\dot{\boldsymbol{x}}}_c^2$, and $\hat{\ddot{\boldsymbol{x}}}_c^2$ as signals of the reference orientation, reference speed, and reference acceleration. The schematic diagram of impedance control method of the exoskeleton suit based on tracking differentiator is as shown in Fig. 6.1. For the exoskeleton suit as the researched object in this book, $\hat{\boldsymbol{x}}_c$ is three-dimensional, so actually each stage of tracking differentiator has three specific tracking differentiators.

For each specific tracking differentiator, in this book we uses the tracking differentiator in discrete form as follows:

$$\begin{cases} x_1(k+1) = x_1(k) + hx_2(k) \\ x_2(k+1) = x_2(k) + hw \end{cases}, \quad |w| \le r \tag{6.21}$$

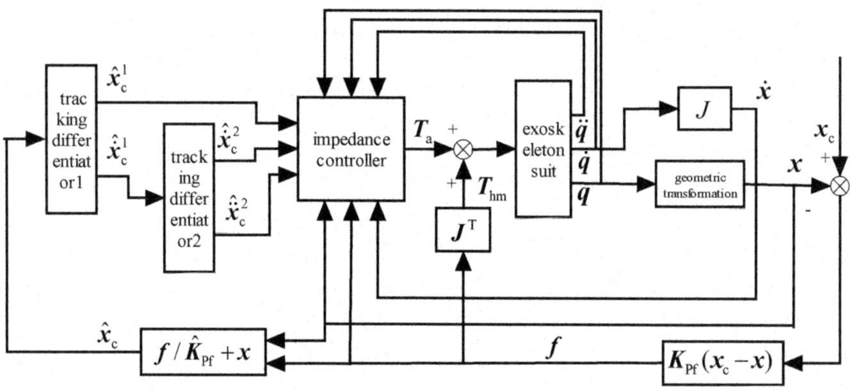

Fig. 6.1 Impedance control schematic diagram of the exoskeleton suit based on the tracking differentiator

$$
\begin{aligned}
w &= \mathrm{fst}(x_1(k) - v(k), x_2(k), r, h) \\
&= -\begin{cases} r\mathrm{sign}(a), & |a| > d \\ r\frac{a}{d}, & |a| \leq d \end{cases} \\
a &= \begin{cases} x_2(k) + \frac{a_0 - d}{2}\mathrm{sign}(y), & |y| > d_0 \\ x_2(k) + \frac{y}{h}, & |y| \leq d_0 \end{cases} \\
a_0 &= \sqrt{d^2 + 8r|y|} \\
y &= x_1(k) - v(k) + hx_2(k) \\
d_0 &= hd \\
d &= rh
\end{aligned}
\tag{6.22}
$$

In the equation, v represents the input of the tracking differentiator; x_1, x_2 represent the output of tracking differentiator; x_1 tracks v, and x_2 is the approximate differential of v; function fst has two adjustable parameters r and h, in which r is the speed factor deciding the speed of tracking; and h is the filter factor with the filtering effect on the noise.

6.4 Simulation

Similar to Sect. 5.4, the controlled object is still the support leg model shown in Fig. 2.5a. Simulate in the M language environment of MATLAB. The human body parameter of Winter is used as the mass and geometric attributes of the exoskeleton suit. The initial orientation and reference orientation of the exoskeleton suit are the same as that in Sect. 5.4.1. And through the same filtering processing, get the smooth reference signal.

The other simulation conditions are set as follows:

1. The human-machine interaction model parameter is set as follows:

$$
\boldsymbol{K}_{\mathrm{Pf}} = \mathrm{diag}[\,10000 \quad 10000 \quad 10000\,] \tag{6.23}
$$

2. The sampling period is set as 1 ms;
3. The desired impedance parameter is as follows:

$$
\boldsymbol{K}_{\mathrm{Mp}} = \mathrm{diag}[\,20 \quad 10 \quad 1\,] \tag{6.24}
$$

$$
\boldsymbol{K}_{\mathrm{Dp}} = \mathrm{diag}[\,2000 \quad 800 \quad 150\,] \tag{6.25}
$$

$$
\boldsymbol{K}_{\mathrm{Pp}} = \mathrm{diag}[\,0 \quad 0 \quad 0\,] \tag{6.26}
$$

6.4.1 Simulation of Impedance Control

Use Eq. (6.19) to estimate the reference trajectory, and assume that the human-machine interaction model parameter can be obtained accurately, namely $\hat{\boldsymbol{K}}_{\mathrm{Pf}} = \boldsymbol{K}_{\mathrm{Pf}}$.

The simulation results are as shown in Figs. 6.2, 6.3, 6.4, 6.5, 6.6, 6.7, 6.8, and 6.9. Figure 6.2 shows the orientation tracking curve of the exoskeleton suit's terminal in operation space. Figure 6.3 shows the corresponding orientation tracking error curve. The two figures illustrate that the designed impedance controller can control the exoskeleton suit to track the desired orientation accurately, and the tracking error is at 10^{-4} orders of magnitude.

Figure 6.4 shows the joint torque exerted by each joint actuator in joint space in the process of trajectory tracking; Fig. 6.5 shows the expression form of the joint torque exerted by the actuator being transformed into operation space; Fig. 6.6 shows the human-machine acting force of the operator exerting on the exoskeleton suit in joint space; Fig. 6.7 shows the acting force of the operator exerting on the

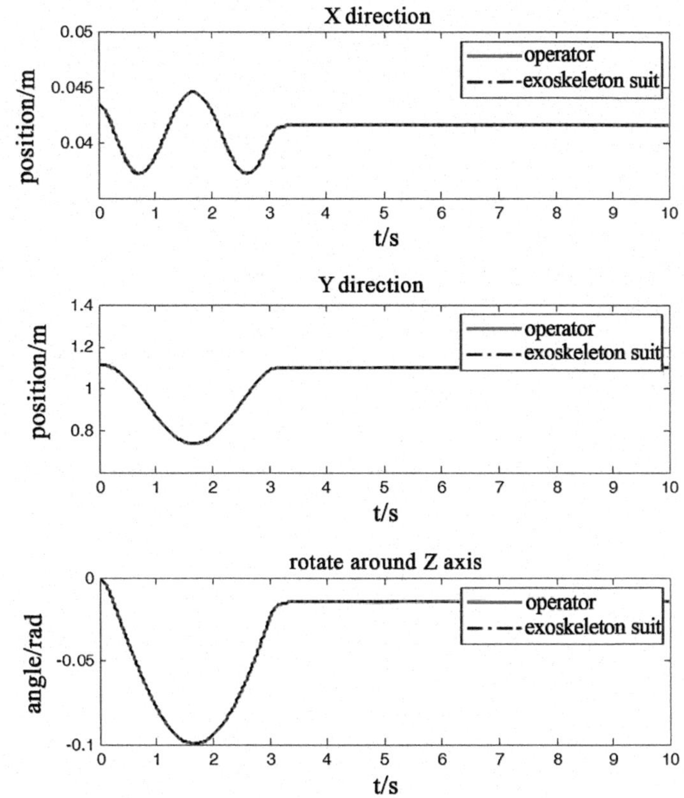

Fig. 6.2 Orientation tracking curve of operation space

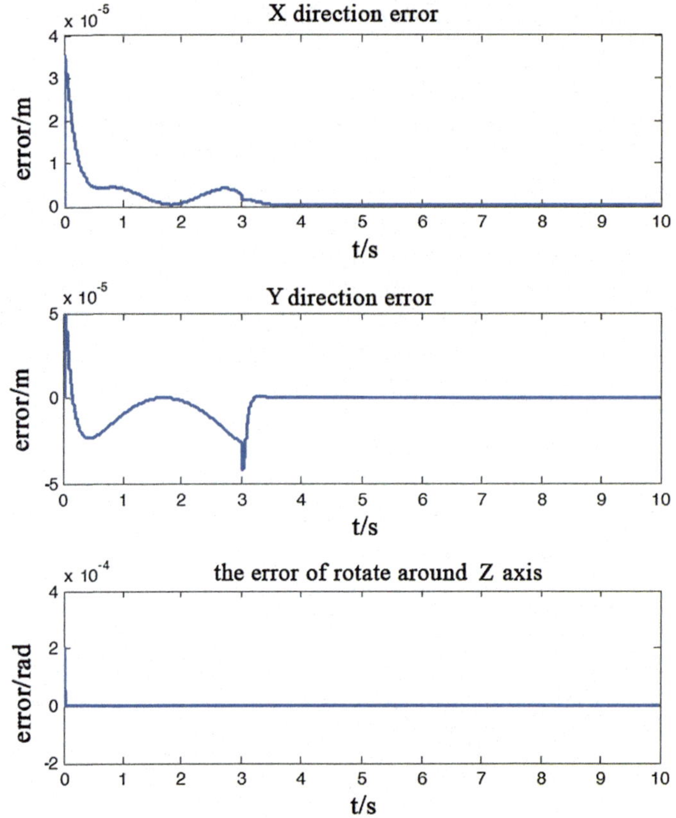

Fig. 6.3 Orientation tracking error curve of operation space

Fig. 6.4 Joint control torque exerted by the actuator

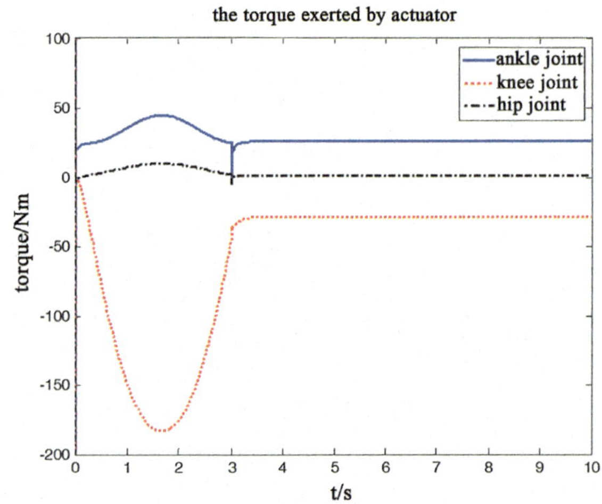

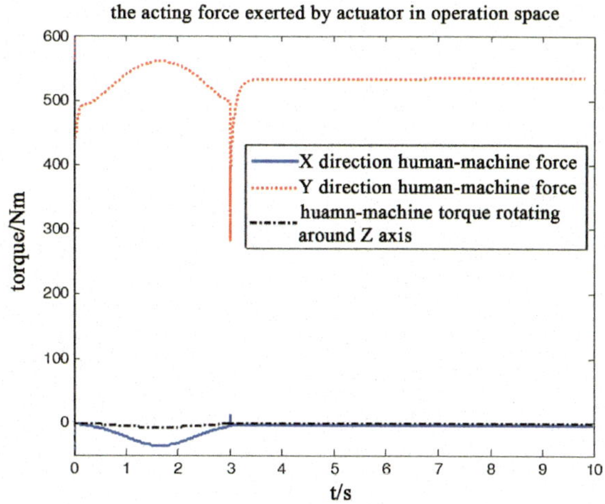

Fig. 6.5 Acting force exerted by the actuator in operation space

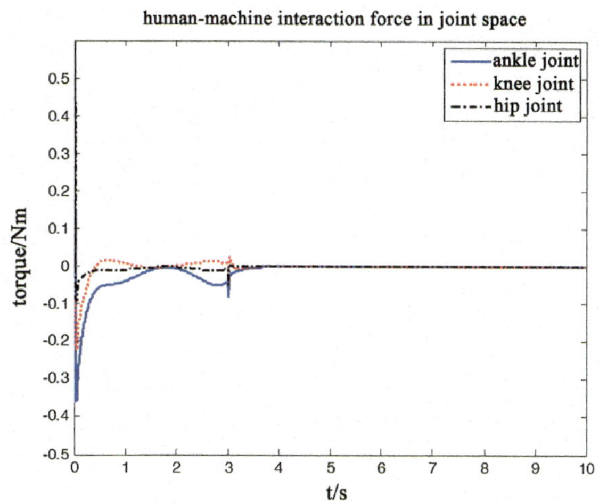

Fig. 6.6 Human-machine interaction force of joint space

exoskeleton suit in operation space. From Figs. 6.4, 6.5, 6.6, and 6.7, we can see that compared with torque exerted by the actuator, the human-machine acting force is very small. Figure 6.8 shows the human factor ***HF*** of joint space and the corresponding maximum human factor $\boldsymbol{HFM} = [0.0055 \quad 0.0060 \quad 0.0064]$; Fig. 6.9 shows the corresponding human factor ***HF*** of operation space and the maximum human factor $\boldsymbol{HFM} = [0.0002 \quad 0.0003 \quad 0.0036]$. From Figs. 6.8 and 6.9, we can see that both in joint space and in operation space, human factor ***HF*** is almost close to zero, and each element of the maximum human factor ***HFM*** is also under 10^{-3} orders of magnitude, illustrating that the proportion of human in exoskeleton suit's movement is also close to zero, the actuator provides the most power of exoskeleton suit's movement, and the energy consumed by human is very small, so the design purpose of the controller has been achieved.

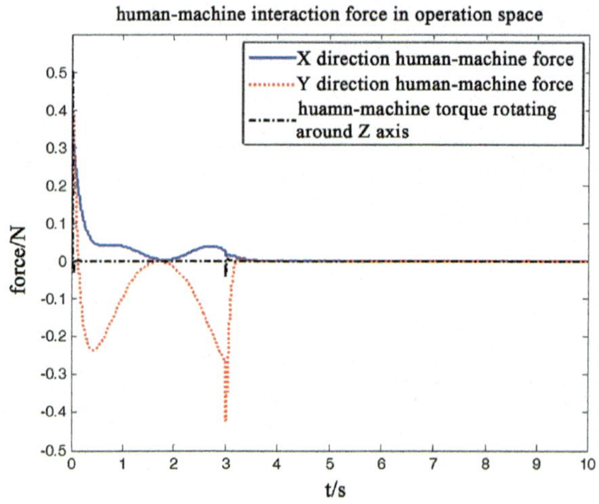

Fig. 6.7 Human-machine interaction force of operation space

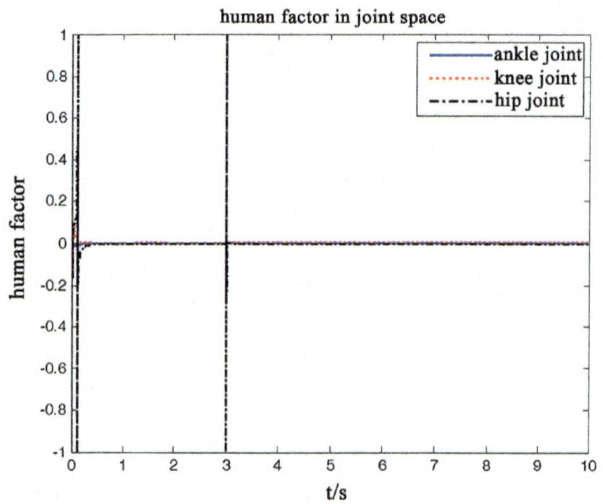

Fig. 6.8 Human factor of joint space

6.4.1.1 The Influence of the Human-Machine Interaction Model Parameter on Human-Machine Force

The previous simulation is made on the basis of the accurate estimation of the human-machine interaction model parameter matrix; however, in fact, it is difficult for us to estimate the parameter matrix accurately. As the description in Sect. 6.2, although we cannot estimate the parameter accurately, as long as the estimated value $\hat{\boldsymbol{K}}_{\mathrm{Pf}}$ is positive definite, we can guarantee the human-machine force $\boldsymbol{f}$ converges to zero. Therefore, we simulate it to verify. We make the simulation in five different conditions: the normal ($\hat{\boldsymbol{K}}_{\mathrm{Pf}} = \boldsymbol{K}_{\mathrm{Pf}}$), the cases of increasing by 20% ($\hat{\boldsymbol{K}}_{\mathrm{Pf}} = 1.2 \times \boldsymbol{K}_{\mathrm{Pf}}$), the cases of increasing by 40% ($\hat{\boldsymbol{K}}_{\mathrm{Pf}} = 1.4 \times \boldsymbol{K}_{\mathrm{Pf}}$), the cases of

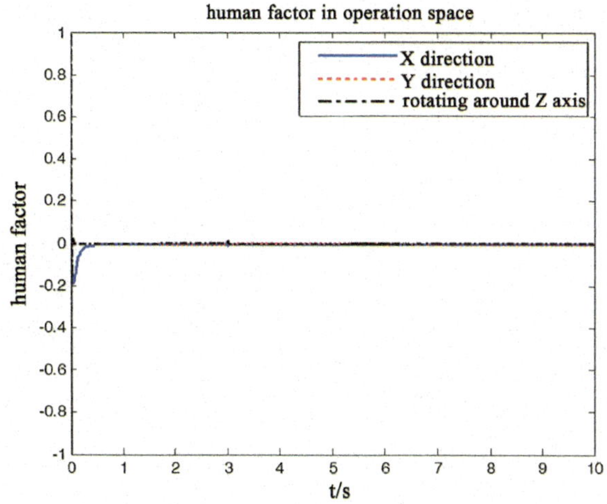

Fig. 6.9 Human factor of operation space

decreasing by 20% ($\hat{\boldsymbol{K}}_{\mathrm{Pf}} = 0.8 \times \boldsymbol{K}_{\mathrm{Pf}}$), and the cases of decreasing by 40% ($\hat{\boldsymbol{K}}_{\mathrm{Pf}} = 0.6 \times \boldsymbol{K}_{\mathrm{Pf}}$).

The simulation results are as shown in Figs. 6.10, 6.11, 6.12, and 6.13. Among them, in Figs. 6.10, 6.11, and 6.12 are shown the human-machine interaction force (torque) of X direction and Y direction and the rotation direction around Z-axis in operation space, and Fig. 6.13 is the partial enlarged drawing of Fig. 6.12 within 0.1 s. From the simulation results, we can see that even though the estimation for the human-machine interaction model parameter $\boldsymbol{K}_{\mathrm{Pf}}$ is very inaccurate, the human-machine interaction force still can be guaranteed within a very small range.

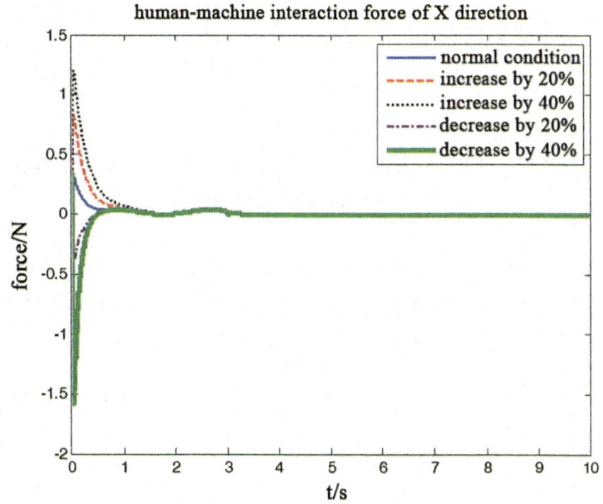

Fig. 6.10 Human-machine interaction force contrast curve in X direction

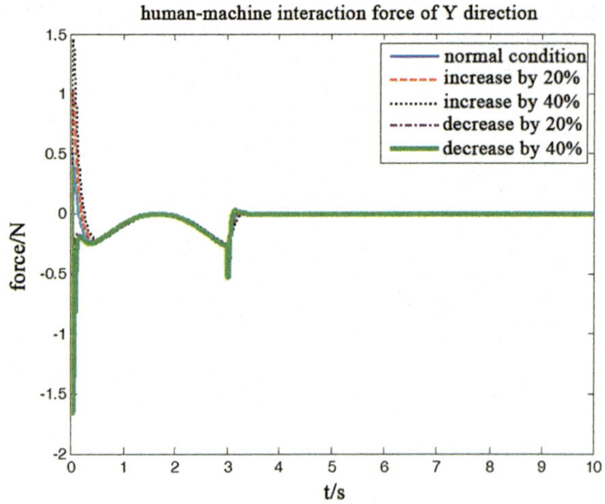

Fig. 6.11 Human-machine interaction force contrast curve in Y direction

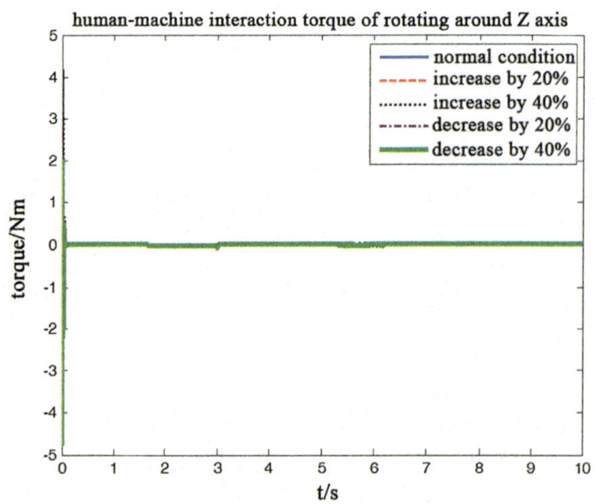

Fig. 6.12 Human-machine interaction torque contrast curve around Z-axis

6.4.1.2 The Influence of Model Parameter Change on Human-Machine Force

In Eq. (6.13), when $\Delta \boldsymbol{A}(\boldsymbol{q}) = \Delta \boldsymbol{B}(\boldsymbol{q}) = \Delta \boldsymbol{D}(\boldsymbol{q}) \neq 0$, the control moment provided by the controller cannot guarantee the trajectory tracking performance. Let $\boldsymbol{f} = \boldsymbol{f}_1 + \boldsymbol{f}_2$, and let

$$-\boldsymbol{f}_1 = \boldsymbol{K}_{\mathrm{Mp}} \Delta \ddot{\boldsymbol{x}} + \boldsymbol{K}_{\mathrm{Dp}} \Delta \dot{\boldsymbol{x}} + \boldsymbol{K}_{\mathrm{Pp}} \Delta \boldsymbol{x} \quad (6.27)$$

Then, by Eq. (6.13), there is

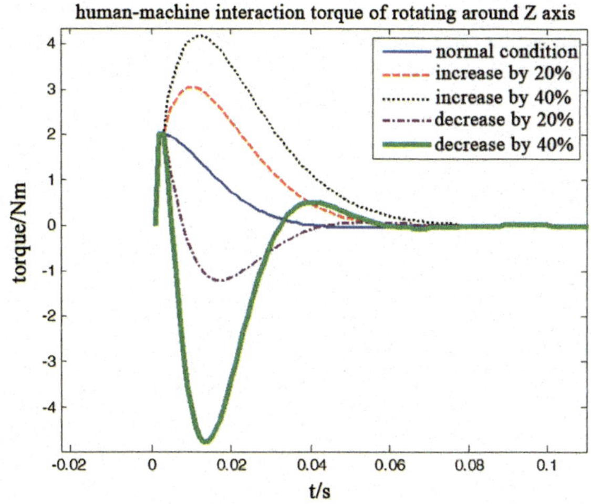

Fig. 6.13 Partial enlarged drawing of Fig. 6.12

$$f_2 = \boldsymbol{K}_{\mathrm{Mp}}\hat{\boldsymbol{A}}(\boldsymbol{q})^{-1}(\Delta\boldsymbol{A}(\boldsymbol{q})\ddot{\boldsymbol{x}} + \Delta\boldsymbol{B}(\boldsymbol{q},\dot{\boldsymbol{q}})\dot{\boldsymbol{x}} + \Delta\boldsymbol{D}(\boldsymbol{q})) \tag{6.28}$$

$\boldsymbol{f}_1$ is used to achieve the desired impedance, while $\boldsymbol{f}_2$ is the extra force exerted by the operator to eliminate the influence of model parameter's uncertain items. In order to observe the magnitude of $\boldsymbol{f}_2$, simulate the change of the model parameter. The previous simulation is to fix the mass of the exoskeleton suit and the load at 70 kg, and the model parameter is calculated also according to 70 kg. In order to reflect the change of model parameter, let the actual mass of the exoskeleton suit and the load be 50 and 90 kg, respectively, and simulate correspondingly; the model parameter in the controller is still calculated according to 70 kg. The simulation results are as shown in Figs. 6.14, 6.15, and 6.16. The simulation results show that the model parameter's change really has great influence on the human-machine interaction force. The change increases the human-machine acting force in the dynamic process, and when in steady state, it does not converge to zero, but stabilize in a constant value. For example, in the figure, after the 3 s, after completing the squatting and rising actions, the human body stands still, while the human-machine acting force still remains at a relatively large constant value; therefore, we had to better find ways to compensate the uncertainty of the system. In Chap. 7, we will use RBF neural network to estimate the uncertainty caused by the model parameter's change and make compensation in the controller to weaken the influence of the uncertainty on human-machine force.

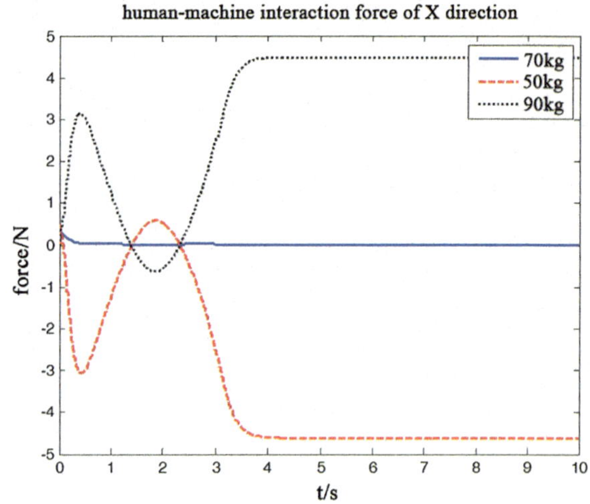

Fig. 6.14 Human-machine interaction force contrast curve in X direction

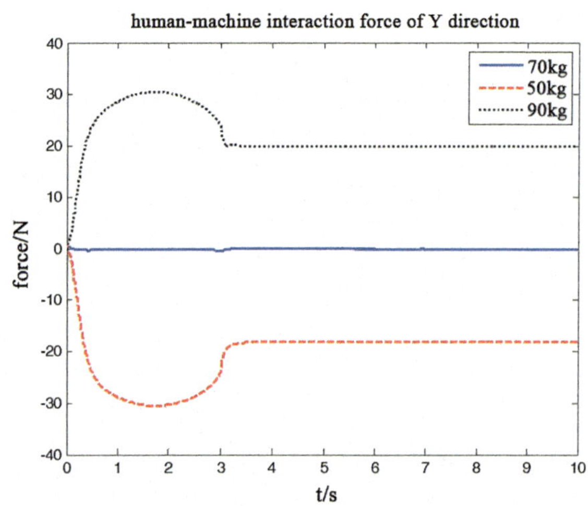

Fig. 6.15 Human-machine interaction force contrast curve in Y direction

6.4.1.3 The Influence of the Measurement Noise on Human-Machine Force

After observing Eq. (6.12), we can see that when making the control moment calculation, $\boldsymbol{x}_c$, $\dot{\boldsymbol{x}}_c$, and $\ddot{\boldsymbol{x}}_c$ need to be used. $\boldsymbol{x}_c$ is generated by the estimation of human-machine force $\boldsymbol{f}$ based on Eq. (6.19), $\dot{\boldsymbol{x}}_c$ is generated by the difference of $\boldsymbol{x}_c$, and $\ddot{\boldsymbol{x}}_c$ is generated by the difference of $\dot{\boldsymbol{x}}_c$. Because $\boldsymbol{f}$ is obtained through the multi-dimensional force/torque sensor's measurement and the measuring process will inevitably introduce measurement noise, so the estimated value of $\boldsymbol{x}_c$ also inevitably contains the measurement noise. Then, when using the finite difference

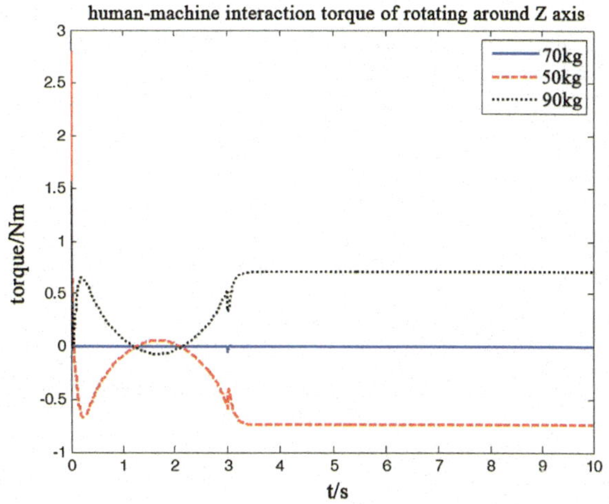

Fig. 6.16 Human-machine interaction torque contrast curve around Z-axis

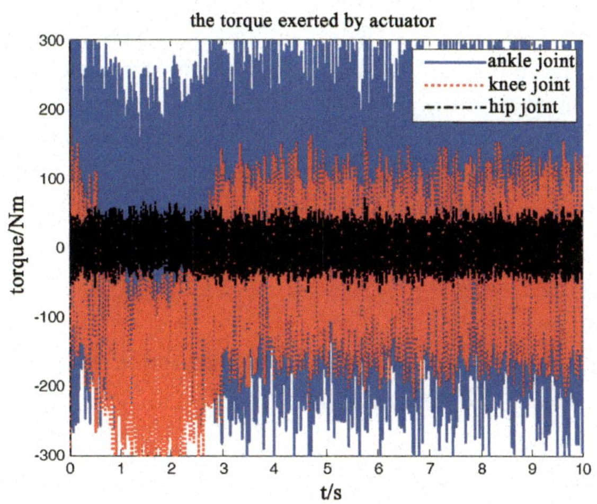

Fig. 6.17 Joint control torque exerted by the actuator containing measurement noise

method to estimate $\dot{\boldsymbol{x}}_c$ and $\ddot{\boldsymbol{x}}_c$, the noise will be inevitably amplified, thus affecting the control effect of the controller, even causing system instability.

In order to observe the influence of measurement noise on the controller, adding random noise signal into the sensor signal (f) to make simulation, signal amplitude being 0.015 N, and the other simulation conditions are same as that in Sect. 6.4.1. The simulation results are as shown in Figs. 6.17 and 6.18. Figure 6.17 shows the joint torque of the actuator output, from which we can see that even if the given noise amplitude is very small, the actuator torque signal generated by the controller's calculation will also amplify the noise, and the output signal of the actuator has huge oscillations in amplitude, which will inevitably shorten the life of the actuator, even damaging the actuator or other devices; therefore, some certain

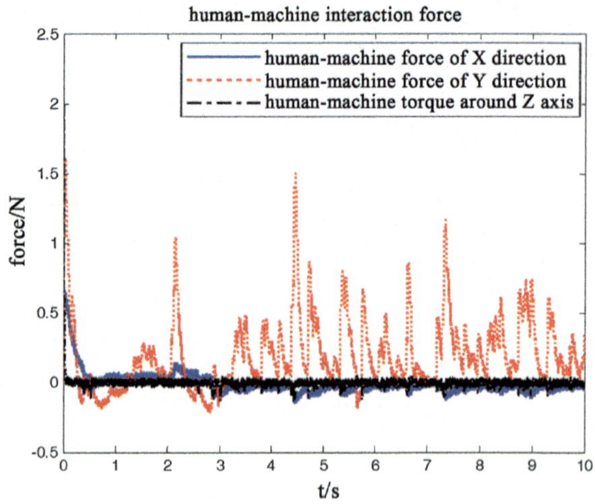

Fig. 6.18 Human-machine interaction force in operation space containing measurement noise

measures must be taken to suppress the influence of noise on the controller. In the next section, we will use tracking differentiator to weaken the noise effect. Figure 6.18 shows the human-machine interaction force containing noise, compared with the control force exerted by the actuator in operation space shown in Fig. 6.5, although the human-machine interaction force has increased smaller, which indicates that the operator's energy consumption has increased not much, still there are many peaks in the curve, which will obviously affect the operator's comfort degree.

6.4.2 Simulation of Impedance Control Based on Tracking Differentiator

The simulation results of Sect. 6.4.1.3 show that the noise can affect the control effect of the controller. Section 6.3 has introduced the method of using the two-stage tracking differentiator to estimate the reference orientation, reference speed, and reference acceleration, and here, we make the simulation to verify. The particular forms of tracking differentiator are as shown in Eqs. (6.21) and (6.22). When simulating, set the parameters of the tracking differentiator: $r = 200$, $h = 0.001$, and set the sensor's noise amplitude as 0.5 N. The desired impedance parameters of impedance controller are $\boldsymbol{K}_{\text{Mp}} = \text{diag}[2 \quad 0.5 \quad 1]$, $\boldsymbol{K}_{\text{Dp}} = \text{diag}[200 \quad 20 \quad 50]$, and $\boldsymbol{K}_{\text{Pp}} = \text{diag}[0 \quad 0 \quad 0]$, and the other simulation conditions are same as those in Sect. 6.4.

Because adding the tracking differentiator is to suppress the effect of measurement noise, therefore, the following simulations are conducted on the basis of containing noise, and compared with Sect. 6.4.1.3, the noise amplitude is increased to 0.5 N. The simulation results are as shown in Figs. 6.19, 6.20, 6.21, and 6.22.

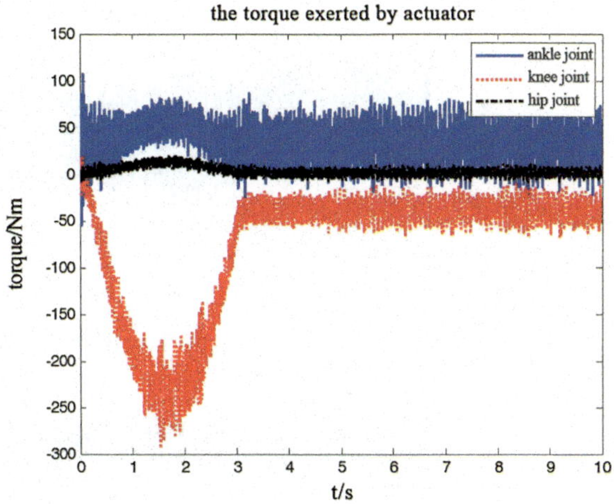

Fig. 6.19 Joint control moment exerted by the actuator

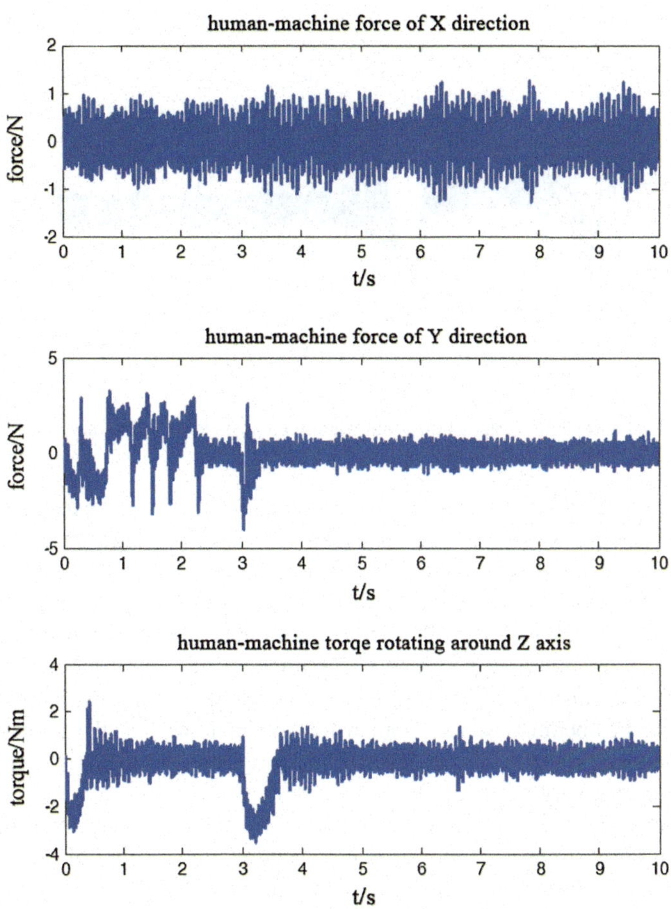

Fig. 6.20 Human-machine interaction force of operation space

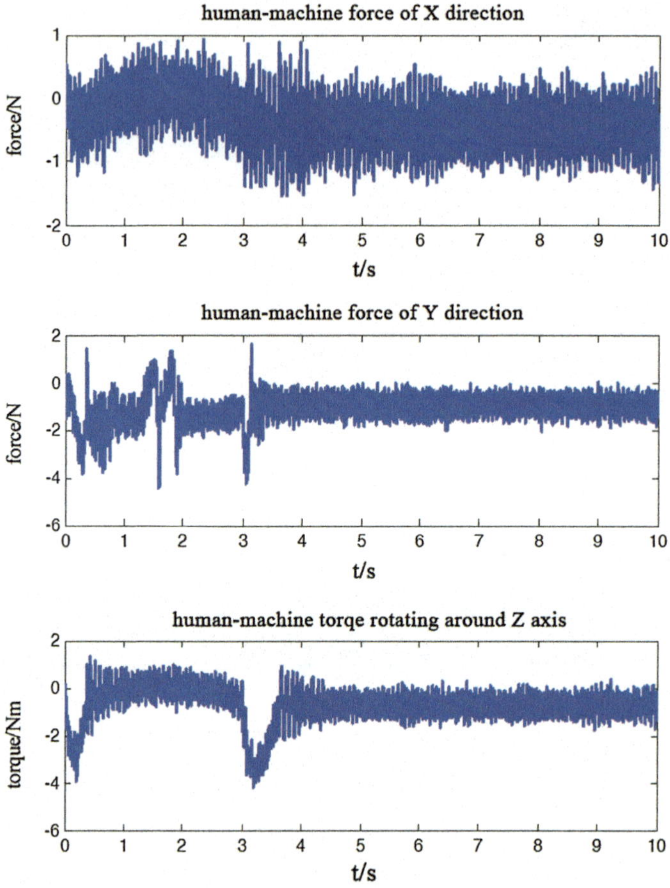

Fig. 6.21 Human-machine interaction force when the exoskeleton suit's mass is 50 kg

The trajectory tracking curve of trunk's center of gravity of the exoskeleton suit in operation space is similar to that in Fig. 6.2, so we no longer give. Figure 6.19 shows the control moment exerted by the actuator in joint space, and the glitch in the curve is caused by the sensor's measurement noise; compared with Fig. 6.17 (the noise amplitude is 0.015 N), even though under the conditions of there being 0.5 N measurement noise, the glitch's amplitude in the control moment exerted by the actuator is much smaller. Figure 6.20 shows the human-machine interaction force curve in operation space, from which we can see that the human-machine interaction force remains at low level, illustrating that the operator's energy consumption is still very small.

Equations (6.19) and (6.20) have illustrated that even though the estimation of the human body stiffness parameter is not accurate, the stability of the system still can be maintained, and the existence of the tracking differentiator does not affect

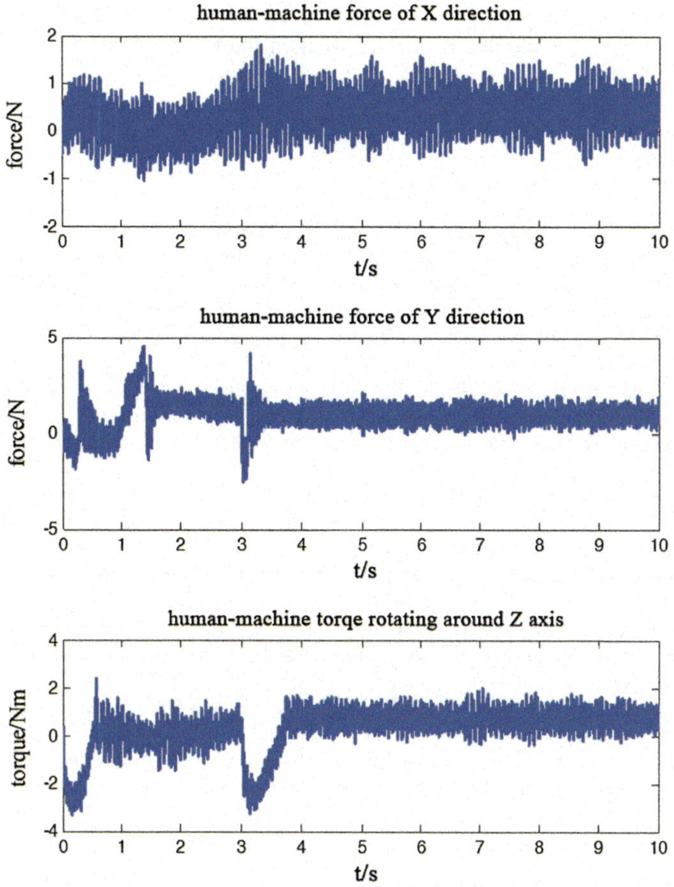

Fig. 6.22 Human-machine interaction force when the exoskeleton suit's mass is 90 kg

this conclusion. Because the simulation results are similar to what shown in Figs. 6.10, 6.11, 6.12, and 6.13, we no longer give here.

Similar to Sect. 6.4.1.2, here study the effect of model parameter's change on system performance, and the simulation results are as shown in Figs. 6.21 and 6.22. Figure 6.21 shows the human-machine interaction force curve when the exoskeleton suit's mass is 50 kg, and Fig. 6.22 shows the human-machine interaction force curve when the exoskeleton suit's mass is 90 kg. Contrasting Figs. 6.20 and 6.21 with Fig. 6.22, we can see that when the model parameter's change exists (namely uncertainty), human-machine interaction force will somewhat increase, thus making the operator's energy consumption increase, same as the conclusion in Sect. 6.4.1.2. And also, we can see that although the tracking differentiator suppresses a part of the noise impact, it is not complete; in the follow-up work or in the practical engineering, the methods of hardware filtering and software filtering also can be adopted to further weaken the effect of noise.

References

1. Yin Yuehong, Wei Zhongxin, Huang Xiaoxi. Force Sensing and Force Control Technology for Intelligent Machine System [M]. Beijing: National Defence Industry Press, 2002.
2. Gabriel, Aguirre-Ollinger, J. Edward Colgate, Michael A. Peshkin, Ambarish Goswami. Active-Impedance Control of a Lower-Limb Assistive Exoskeleton [C]. Proceedings of the 2007 IEEE 10th International Conference on Rehabilitation Robotics Noordwijk, The Netherlands, 2007: 188–195.
3. Heng, Wang, K. H. Low, Michael Yu Wang. Reference Trajectory Generation for Force Tracking Impedance Control by Using Neural Network-based Environment Estimation [C]. 2006 IEEE Conference on Robotics Automation and Mechatronics, Bangkok, Thailand, 2006: 1–6.
4. Qiao, Bing, Lu Rongjian. Impedance force control for position controlled robotic manipulators under the constraint of unknown environments [J]. Journal of Southeast University (English Edition), 2003, 19 (4): 359–363.
5. Chen Feng, Fei Yanqiong, Zhao Xifang. The Impedance Control Methods for Robots [J]. Control and Detection, 2005, (12): 46–48.
6. Wang Yaonan. Robot Intelligent Control Engineering [M]. Beijing: Science Press, 2004.
7. Siciliano, Bruno, Luigi Villani. Robot Force Control [M]. Boston/Dordrecht/London: Kluwer Academic Publishers, 1999.
8. Han Jingqing, Wang Wei. Nonlinear Tracking Differentiator [J]. Journal of Systems Science and Mathematical Sciences, 1994, 14 (2): 177–183.
9. Han Jingqing, Yuan Lulin. The Discrete Form of Tracking Differentiator [J]. Journal of Systems Science and Mathematical Sciences, 1999, 19 (6): 268–273.

Chapter 7
Impedance Control of Exoskeleton Suit with Uncertainties

7.1 Introduction

The exoskeleton suit is a very complex multiple-input–multiple-output system with time-varying, strong coupling, and nonlinear dynamical characteristics. Because of the inaccurate measurement and modeling, the change of the load, and the influence of external disturbance, actually the accurate complete motion model of the exoskeleton suit cannot be obtained, which is so-called uncertainty. The uncertainty can be divided into the unknown parameter uncertainty, structural uncertainty, and external disturbance uncertainty [1, 2]. In this chapter, we only consider parameter uncertainty.

The simulation results of the previous chapter show that when using the impedance control method to control the exoskeleton suit, if the load borne by human changes, then the acting force of human exerting on the exoskeleton suit will increase, which will cause the increase of human body's energy expenditure. And the load can be regarded as a part of the weight of the exoskeleton suit's trunk; so when the load changes, the weight of the exoskeleton suit's trunk and the moment of inertia will change, which are the important reflection of the exoskeleton suit's parameter uncertainty. Many scholars have put forward many kinds of control methods for the system with uncertain factors, such as neural network control, fuzzy control, variable structure control, and robust control. [1, 3–10]. In this chapter, based on the impedance control method of the exoskeleton suit introduced in the previous chapter, we will use RBF neural network with the adaptive learning algorithm to estimate the uncertainty of the exoskeleton suit system, and make compensation, thus effectively reducing the effect of uncertainty on the human-machine acting force, and reducing the energy consumption of the human body.

Z. Yang et al., *Force Control Theory and Method of Human Load Carrying Exoskeleton Suit*, DOI 10.1007/978-3-662-54144-9_7

7.2 Estimation of Neural Networks with Uncertainties

From Eq. (6.5), we have got the dynamic equation of the exoskeleton suit in operation space:

$$\boldsymbol{A}(\boldsymbol{q})\ddot{\boldsymbol{x}} + \boldsymbol{B}(\boldsymbol{q}, \dot{\boldsymbol{q}})\dot{\boldsymbol{x}} + \boldsymbol{D}(\boldsymbol{q}) = \boldsymbol{u} + \boldsymbol{f} \tag{7.1}$$

Based on the inverse dynamic method, design the controller

$$\boldsymbol{u} = \hat{\boldsymbol{A}}(\boldsymbol{q})\boldsymbol{\tau} + \hat{\boldsymbol{B}}(\boldsymbol{q}, \dot{\boldsymbol{q}})\dot{\boldsymbol{x}} + \hat{\boldsymbol{D}}(\boldsymbol{q}) - \boldsymbol{f} \tag{7.2}$$

$$\boldsymbol{\tau} = \ddot{\boldsymbol{x}}_{\mathrm{c}} + \boldsymbol{K}_{\mathrm{Mp}}{}^{-1}(\boldsymbol{K}_{\mathrm{Dp}}\Delta\dot{\boldsymbol{x}} + \boldsymbol{K}_{\mathrm{Pp}}\Delta\boldsymbol{x} + \boldsymbol{f}) \tag{7.3}$$

After substituting Eqs. (7.2) and (7.3) into Eq. (7.1), we get the dynamic equation of the closed-loop system

$$\Delta\ddot{\boldsymbol{x}} + \boldsymbol{K}_{\mathrm{Mp}}{}^{-1}(\boldsymbol{K}_{\mathrm{Dp}}\Delta\dot{\boldsymbol{x}} + \boldsymbol{K}_{\mathrm{Pp}}\Delta\boldsymbol{x} + \boldsymbol{f}) = \hat{\boldsymbol{A}}(\boldsymbol{q})^{-1}(\Delta\boldsymbol{A}(\boldsymbol{q})\ddot{\boldsymbol{x}} + \Delta\boldsymbol{B}(\boldsymbol{q}, \dot{\boldsymbol{q}})\dot{\boldsymbol{x}} + \Delta\boldsymbol{D}(\boldsymbol{q})) \tag{7.4}$$

In ideal conditions $\Delta\boldsymbol{A}(\boldsymbol{q}) = \Delta\boldsymbol{B}(\boldsymbol{q}) = \Delta\boldsymbol{D}(\boldsymbol{q}) = 0$, the behavior of human-machine force and the exoskeleton suit meets the target impedance relationship

$$-\boldsymbol{f} = \boldsymbol{K}_{\mathrm{Mp}}\Delta\ddot{\boldsymbol{x}} + \boldsymbol{K}_{\mathrm{Dp}}\Delta\dot{\boldsymbol{x}} + \boldsymbol{K}_{\mathrm{Pp}}\Delta\boldsymbol{x} \tag{7.5}$$

The above is the exoskeleton suit's impedance control; however, in the actual system, the uncertainty of the exoskeleton suit system exists inevitably, namely $\Delta\boldsymbol{A}(\boldsymbol{q}) \neq 0$, $\Delta\boldsymbol{B}(\boldsymbol{q}, \dot{\boldsymbol{q}}) \neq 0$, $\Delta\boldsymbol{D}(\boldsymbol{q}) \neq 0$. At this time, the control moment provided by the controller cannot guarantee the trajectory tracking performance, and the operator needs to exert extra moment to eliminate the influence of model parameter's uncertainty, increasing the energy consumption of the operator; for the specific detail, please refer to simulation results of the last chapter; therefore, the model's uncertain items must be estimated and compensated.

Write the system equation as the state-space equation. Assume $\boldsymbol{z} = [\Delta\boldsymbol{x} \quad \Delta\dot{\boldsymbol{x}}]^{\mathrm{T}}$, and from Eq. (6.18),we can know $\boldsymbol{f}$ and the system state $\boldsymbol{z}$ satisfy Eq. (7.6)

$$\boldsymbol{f} = [\boldsymbol{K}_{\mathrm{pf}} \quad 0]\boldsymbol{z} \tag{7.6}$$

At the same time, let the uncertain items of the system dynamics be

$$\boldsymbol{\Psi} = \hat{\boldsymbol{A}}(\boldsymbol{q})^{-1}[\Delta\boldsymbol{A}(\boldsymbol{q})\ddot{\boldsymbol{p}} + \Delta\boldsymbol{B}(\boldsymbol{q}, \dot{\boldsymbol{q}})\dot{\boldsymbol{p}} + \Delta\boldsymbol{D}(\boldsymbol{q})] \tag{7.7}$$

Then, under the function of control law (7.2), Eq. (7.3), the error state equation of the system is

$$\dot{z} = \boldsymbol{M}z + \boldsymbol{N}\boldsymbol{\Psi} \tag{7.8}$$

In the equation

$$\boldsymbol{M} = \begin{bmatrix} 0 & \boldsymbol{I} \\ -(\boldsymbol{K}_{\mathrm{Pp}} + \boldsymbol{K}_{\mathrm{pf}})/\boldsymbol{K}_{\mathrm{Mp}} & -\boldsymbol{K}_{\mathrm{Dp}}/\boldsymbol{K}_{\mathrm{Mp}} \end{bmatrix} \tag{7.9}$$

$$\boldsymbol{N} = \begin{bmatrix} 0 \\ \boldsymbol{I} \end{bmatrix} \tag{7.10}$$

Assume that the uncertain items of the system are known and then redesign the control law Eq. (7.3) to be

$$\boldsymbol{\tau} = \ddot{\boldsymbol{x}}_{\mathrm{c}} + \boldsymbol{K}_{\mathrm{Mp}}{}^{-1}(\boldsymbol{K}_{\mathrm{Dp}}\Delta\dot{\boldsymbol{x}} + \boldsymbol{K}_{\mathrm{Pp}}\Delta\boldsymbol{x} + \boldsymbol{f}) + \boldsymbol{\Psi} \tag{7.11}$$

After substituting the control law Eqs. (7.2) and (7.11) into the system Eq. (7.1), we get the desired equation of the closed-loop system (7.5).

However, in the practical engineering, the model uncertain items $\boldsymbol{\Psi}$ are unknown; therefore, it is necessary to identify the uncertain items $\boldsymbol{\Psi}$ to realize the compensation for the uncertain items in control law.

Use RBF neural network to self-adaptively approximate the uncertain items. For the RBF neural network's simple introduction, please refer to Chap. 8. RBF network algorithm is [11]

$$h_i = f(\|z - \varphi_i\|/\sigma_i^2), \quad i = 1, 2, \ldots, m \tag{7.12}$$

$$\boldsymbol{s} = \boldsymbol{\theta}^{\mathrm{T}}\boldsymbol{h}(z) \tag{7.13}$$

In the equation z represents the network's input signal; $\boldsymbol{h}(z) = [h_1, h_2, \ldots h_m]$ represents the output of the gaussian basis function; $\boldsymbol{\theta}$ represents the weights of the neural network; m represents the number of neurons.

Assume

1. the output of the neural network $\boldsymbol{s}(z, \boldsymbol{\theta})$ is continuous;
2. the output of neural network $\boldsymbol{s}(z, \boldsymbol{\theta}^*)$ approximates the continuous function $\boldsymbol{\Psi}$, and there is a very small positive number ε. There is

$$\max\|\boldsymbol{s}(z, \boldsymbol{\theta}^*) - \boldsymbol{\Psi}\| \leq \varepsilon \tag{7.14}$$

In the equation $\boldsymbol{\theta}^*$ is the $n \times n$ matrix, representing neural network weights of the best identification for $\boldsymbol{\Psi}$.

Let $\boldsymbol{\eta} = \boldsymbol{\Psi} - \boldsymbol{s}(z, \boldsymbol{\theta}^*)$ be the neural network modeling error, and the error is bounded, namely

$$\boldsymbol{\eta}_0 = \sup\|\boldsymbol{\Psi} - \boldsymbol{s}(z, \boldsymbol{\theta}^*)\| \tag{7.15}$$

Then, the system equation (7.8) can be written as

$$\dot{z} = \boldsymbol{M}z + \boldsymbol{N}(\boldsymbol{s}(z, \boldsymbol{\theta}^*) + \boldsymbol{\eta}) \tag{7.16}$$

Assume $\hat{\boldsymbol{\theta}}$ is the estimated value of $\boldsymbol{\theta}^*$, and according to Eq. (7.11), using the neural network to estimate the uncertain items of the system model, we get the new control law

$$\boldsymbol{\tau} = \ddot{\boldsymbol{x}}_{\mathrm{c}} + \boldsymbol{K}_{\mathrm{Mp}}{}^{-1}(\boldsymbol{K}_{\mathrm{Dp}}\Delta\dot{\boldsymbol{x}} + \boldsymbol{K}_{\mathrm{Pp}}\Delta\boldsymbol{x} + \boldsymbol{f}) + \boldsymbol{s}(z, \hat{\boldsymbol{\theta}}) \tag{7.17}$$

7.3 Stability Analysis

After substituting Eqs. (7.2) and (7.17) into Eq. (7.1), we get

$$\begin{aligned} \dot{z} &= \boldsymbol{M}z + \boldsymbol{N}(\boldsymbol{s}(z, \boldsymbol{\theta}^*) + \boldsymbol{\eta} - \boldsymbol{s}(z, \hat{\boldsymbol{\theta}})) \\ &= \boldsymbol{M}z + \boldsymbol{N}(-\tilde{\boldsymbol{\theta}}\boldsymbol{h}(z) + \boldsymbol{\eta}) \end{aligned} \tag{7.18}$$

In the equation $\tilde{\boldsymbol{\theta}} = \hat{\boldsymbol{\theta}} - \boldsymbol{\theta}^*$.

Define Lyapunov function as

$$V = \frac{1}{2}z^{\mathrm{T}}\boldsymbol{P}z + \frac{1}{2\gamma}\|\tilde{\boldsymbol{\theta}}\|^2 \tag{7.19}$$

In the equation $\gamma > 0$; $\|\tilde{\boldsymbol{\theta}}\|^2 = \mathrm{tr}(\tilde{\boldsymbol{\theta}}^{\mathrm{T}}\tilde{\boldsymbol{\theta}})$, $\mathrm{tr}(\cdot)$ represents the trace of a matrix; matrix $\boldsymbol{P}$ is symmetric positive definite matrix and satisfies the following Lyapunov equation

$$\boldsymbol{PM} + \boldsymbol{M}^{\mathrm{T}}\boldsymbol{P} = -\boldsymbol{Q} \tag{7.20}$$

By differentiating Eq. (7.19), we get

$$\begin{aligned} \dot{V} &= \frac{1}{2}[z^{\mathrm{T}}\boldsymbol{P}\dot{z} + \dot{z}^{\mathrm{T}}\boldsymbol{P}z] + \frac{1}{\gamma}\mathrm{tr}(\dot{\tilde{\boldsymbol{\theta}}}^{\mathrm{T}}\tilde{\boldsymbol{\theta}}) \\ &= -\frac{1}{2}z^{\mathrm{T}}Qz - \boldsymbol{h}^{\mathrm{T}}(z)\tilde{\boldsymbol{\theta}}N^{\mathrm{T}}\boldsymbol{P}z + \boldsymbol{\eta}^{\mathrm{T}}\boldsymbol{N}^{\mathrm{T}}\boldsymbol{P}z + \frac{1}{\gamma}\mathrm{tr}(\dot{\tilde{\boldsymbol{\theta}}}^{\mathrm{T}}\tilde{\boldsymbol{\theta}}) \end{aligned} \tag{7.21}$$

And because

$$\boldsymbol{h}^{\mathrm{T}}(z)\tilde{\boldsymbol{\theta}}\boldsymbol{N}^{\mathrm{T}}\boldsymbol{P}z = \mathrm{tr}[\boldsymbol{N}^{\mathrm{T}}\boldsymbol{P}z\boldsymbol{h}^{\mathrm{T}}(z)\tilde{\boldsymbol{\theta}}] \tag{7.22}$$

Then, Eq. (7.21) can be turned into

$$\dot{V} = -\frac{1}{2}z^{\mathrm{T}}\boldsymbol{Q}z + \boldsymbol{\eta}^{\mathrm{T}}\boldsymbol{N}^{\mathrm{T}}\boldsymbol{P}z + \frac{1}{\gamma}\mathrm{tr}(-\gamma\boldsymbol{N}^{\mathrm{T}}\boldsymbol{P}z\boldsymbol{h}^{\mathrm{T}}(z)\tilde{\boldsymbol{\theta}} + \dot{\tilde{\boldsymbol{\theta}}}^{\mathrm{T}}\tilde{\boldsymbol{\theta}}) \tag{7.23}$$

Therefore, if we adopt the self-adaptive law [11]

$$\dot{\tilde{\boldsymbol{\theta}}}^{\mathrm{T}} = \gamma\boldsymbol{N}^{\mathrm{T}}\boldsymbol{P}z\boldsymbol{h}^{\mathrm{T}}(z) \tag{7.24}$$

Namely

$$\dot{\tilde{\boldsymbol{\theta}}} = \gamma\boldsymbol{h}(z)z^{\mathrm{T}}\boldsymbol{P}\boldsymbol{N} \tag{7.25}$$

Then,

$$\dot{V} = -\frac{1}{2}z^{\mathrm{T}}Qz + \boldsymbol{\eta}^{\mathrm{T}}\boldsymbol{N}^{\mathrm{T}}\boldsymbol{P}z \tag{7.26}$$

From the known number

$$\|\boldsymbol{\eta}^{\mathrm{T}}\| \leq \|\boldsymbol{\eta}_0\|, \quad \|\boldsymbol{N}\| = 1$$

And assume $\lambda_{\min}(\boldsymbol{Q})$ is the proper value's minimum value of matrix $\boldsymbol{Q}$, and assume $\lambda_{\max}(\boldsymbol{P})$ is the proper value's maximum value of matrix $\boldsymbol{P}$, then

$$\begin{aligned}\dot{V} &\leq -\frac{1}{2}\lambda_{\min}(\boldsymbol{Q})\|z\|^2 + \|\boldsymbol{\eta}_0\|\lambda_{\max}(\boldsymbol{P})\|z\| \\ &= -\frac{1}{2}\|z\|[\lambda_{\min}(\boldsymbol{Q})\|z\| - 2\|\boldsymbol{\eta}_0\|\lambda_{\max}(\boldsymbol{P})]\end{aligned} \tag{7.27}$$

So, to make $\dot{V} \leq 0$, we need

$$\lambda_{\min}(\boldsymbol{Q}) \geq \frac{2\|\boldsymbol{\eta}_0\|\lambda_{\max}(\boldsymbol{P})}{\|z\|} \tag{7.28}$$

Then, the radius of convergence of z is

$$\|z\| = \frac{2\|\boldsymbol{\eta}_0\|\lambda_{\max}(\boldsymbol{P})}{\lambda_{\min}(\boldsymbol{Q})} \tag{7.29}$$

So, when the proper value of $\boldsymbol{Q}$ is larger, the proper value of $\boldsymbol{P}$ will be smaller, the upper bound $\boldsymbol{\eta}_0$ of the modeling error of the neural network will be smaller, then the radius of convergence of z will be smaller, the tracking effect will be better, human-machine force also will be smaller, the human body's energy consumption also will be smaller.

7.4 Simulation

Use the exoskeleton suit of single degree of freedom as the object for simulation, and the schematic diagram of the exoskeleton suit of single degree of freedom is as shown in Fig. 3.1. Assume the human leg and the hip joint of the exoskeleton suit are fixed together, connected through a multi-dimensional force/torque sensor on the ankle joint; only considering the movement in the vertical plane, then we can use only one dimension of the sensor, that is, used to measure the dimension of T_{hm}; because it is single degree of freedom, the system's joint space and the operation space coincide. At this time, the dynamic equation of the exoskeleton suit system is as follows:

$$H_0\ddot{q} + C_0\dot{q} + G_0 = T_{\mathrm{a}} + T_{\mathrm{he}} \tag{7.30}$$

In the equation $H_0 = ml^2/3$; $G_0 = mgl\cos(q)$; C_0 is the friction coefficient; q_{c} represents human's movement trajectory and set $q_{\mathrm{c}} = \sin 2\pi t$ (here, q_{c} is a one-dimensional vector, that is, the desired movement trajectory of the exoskeleton suit (reference trajectory), equivalent to the reference trajectory vector $\boldsymbol{x}_{\mathrm{c}}$ of operation space). Assume m = 10, l = 1, C_0 = 0.02, K_{Pf} = 500, and the desired impedance parameters are as follows: K_{Mp} = 1, K_{Dp} = 100, K_{Pp} = 10 ($\boldsymbol{K}_{\mathrm{Mp}}$, $\boldsymbol{K}_{\mathrm{Dp}}$, and $\boldsymbol{K}_{\mathrm{Pp}}$—all can be referred to as the impedance parameters, and here, $\boldsymbol{K}_{\mathrm{Mp}}$, $\boldsymbol{K}_{\mathrm{Dp}}$, and $\boldsymbol{K}_{\mathrm{Pp}}$—all are one-dimensional). Design the self-adaptive parameter γ = 1000, matrix $\boldsymbol{Q} = [50 \quad 0; \quad 0 \quad 50]$.

From Eq. (7.9), we can get

$$\boldsymbol{M} = \begin{bmatrix} 0 & 1 \\ -510 & -100 \end{bmatrix} \tag{7.31}$$

Therefore, from Eq. (7.20), we can get the solution

$$\boldsymbol{P} = \begin{bmatrix} 132.652 & 0.049 \\ 0.049 & 0.2505 \end{bmatrix} \tag{7.32}$$

From Eq. (7.32), we can get the proper value $\lambda_{\mathrm{P1}} = 132.652$, $\lambda_{\mathrm{P2}} = 0.2505$, and also we can get the proper value of $\boldsymbol{Q}$, that is, $\lambda_{\mathrm{Q1}} = \lambda_{\mathrm{Q2}} = 50$, so $\lambda_{\max}(\boldsymbol{P}) = 132.652$, $\lambda_{\min}(\boldsymbol{Q}) = 50$; therefore, assume that the upper bound of neural network modeling error $\|\boldsymbol{\eta}_0\| = 0.1$, and from Eq. (7.29), we can get the system's radius of convergence $\|z\| = 0.53$, here, $z = [\Delta q \quad \Delta\dot{q}]^{\mathrm{T}}$. So, under the condition of satisfying the stability conditional expression, the system's radius of convergence is very small.

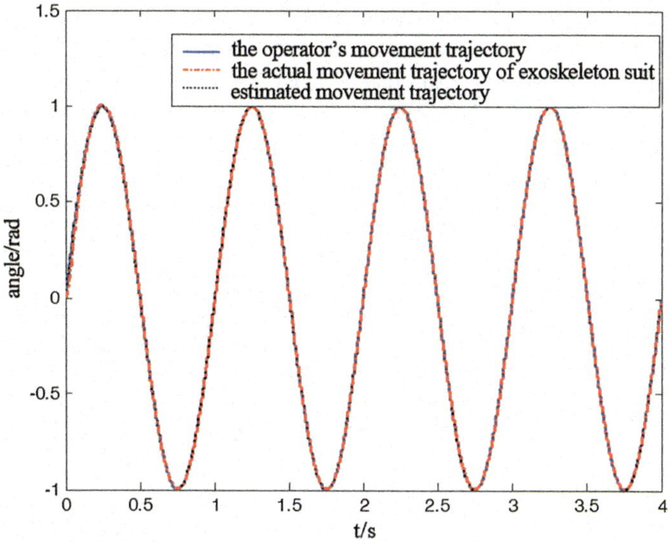

Fig. 7.1 Trajectory tracking curve

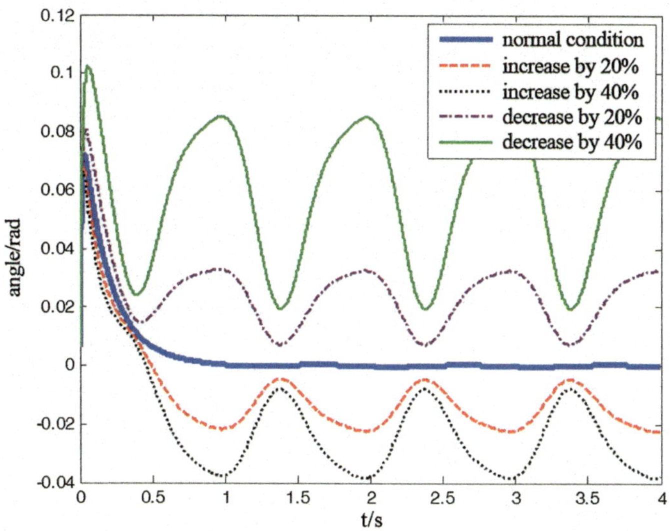

Fig. 7.2 Trajectory tracking error curve

In order to analyze the performance of the neural network compensation, first simulate when without using the neural network compensation, the simulation results as shown in Figs. 7.1, 7.2, 7.3, and 7.4. Figure 7.1 shows the trajectory tracking curve without existing uncertain items: When there are uncertain items, the system's tracking curves are similar, but according to the graphic method of Fig. 7.1, the magnitude of the tracking error cannot be seen obviously, so draw the tracking error curves under different conditions separately for comparison, as shown in Fig. 7.2. From Fig. 7.2, we can see that with the increase of uncertain items, the system's trajectory tracking error keeps increasing, maximally up to 0.1 rad.

Figure 7.3 shows the changing curve of the control moment exerted by the actuator in all circumstances of uncertain change, and Fig. 7.4 shows the changing curve of human-machine acting force in all circumstances of uncertain change. From the two figures, we can see that as the uncertainty changes, the control moment exerted by the actuator also changes, but the absolute value of

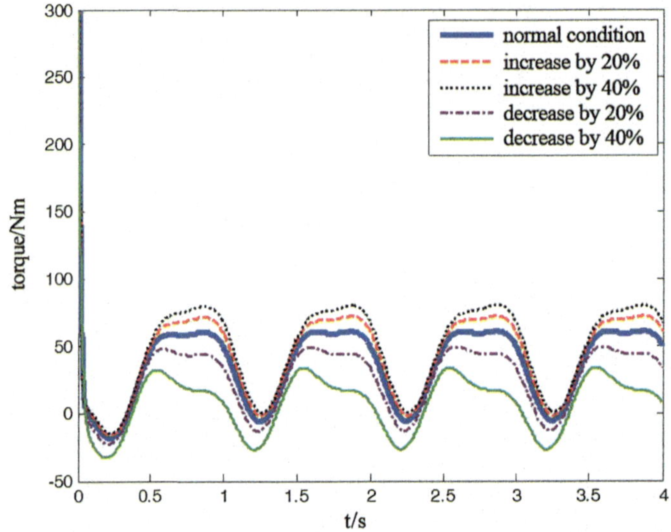

Fig. 7.3 Contrast curve of the control moment exerted by the actuator

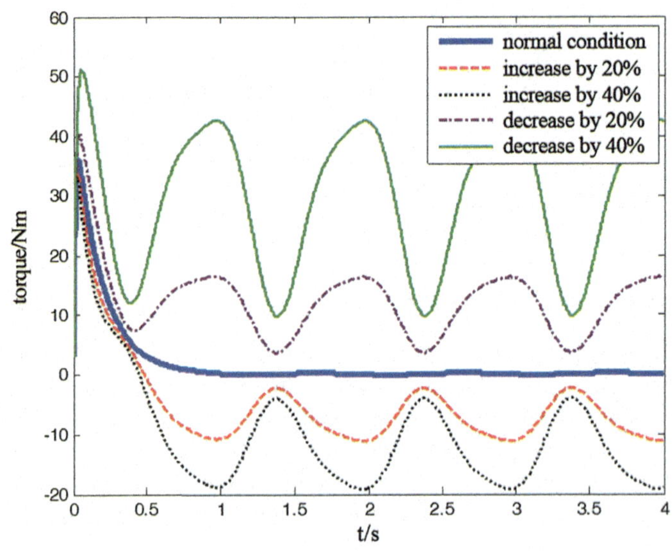

Fig. 7.4 Contrast curve of the human-machine acting force

human-machine acting force increases continuously, and the maximum value can reach 50 Nm, illustrating that the operator's energy consumption increases. The conclusion is the same as that in Sect. 6.4.1.2.

Using the neural network to compensate the uncertainty can improve the performance of the impedance controller and improve its robustness for the system's uncertainty. The simulation results are as shown from Figs. 7.5, 7.6, 7.7, 7.8, 7.9, 7.10, 7.11, 7.12, and 7.13.

Figure 7.5 shows the trajectory tracking error curve of five cases, and the steady-state tracking error is limited in $[-0.02 \quad 0.02]$; the steady-state tracking error without the compensation of neural network shown in Fig. 7.2 is $[-0.04 \quad 0.08]$; therefore, after using the neural network to compensate, the tracking error of the system is also reduced a lot.

Figures 7.6 and 7.7 show the contrast curve of control moment exerted by the actuator in the five cases; compared with Fig. 7.3, the output volatility's amplitude of the actuator decreases, illustrating that the neural network's compensating moment has played a big role.

Figure 7.8 shows the contrast curve of human-machine acting force in the five cases; when there is 40% error in the model parameter, the human-machine acting force is limited in $[-10 \quad 10]$, and when there is 40% error in the model parameter, the human-machine acting force without the compensation of neural network as shown in Fig. 7.4 is within $[-20 \quad 40]$; therefore, we can see that after using the neural networks to compensate, the human-machine acting force decreases obviously, suppressing the influence of uncertainty on the human-machine acting force.

Figs. 7.9, 7.10, 7.11, 7.12, and 7.13 show the estimation of the neural network for model uncertainty in five cases, respectively. From the figures, we can see that

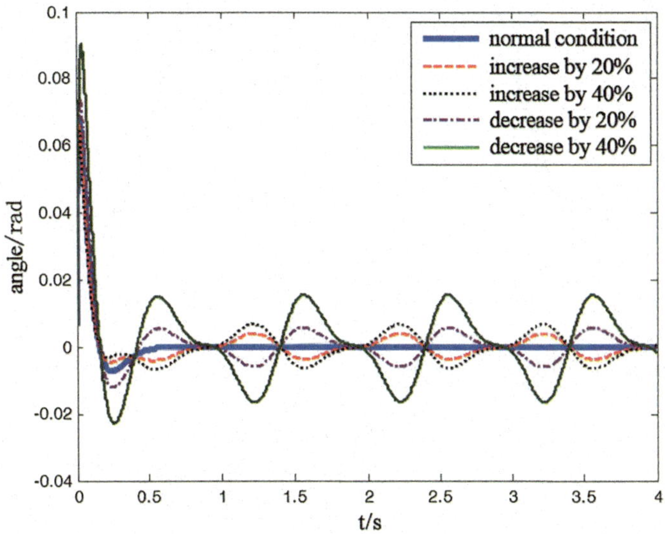

Fig. 7.5 Trajectory tracking error curve

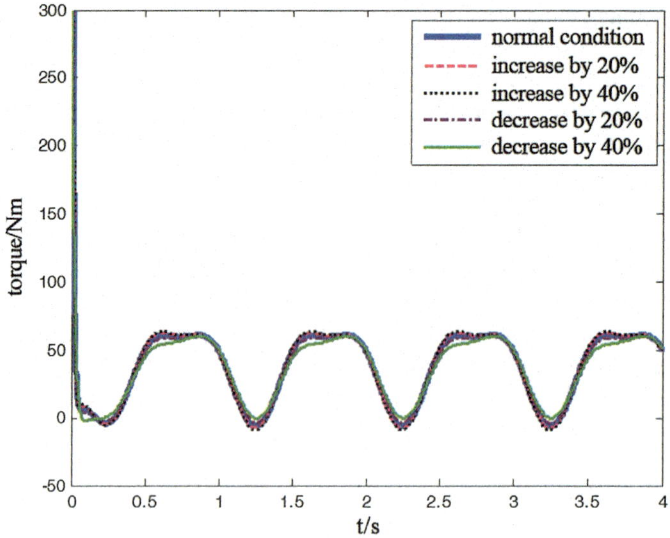

Fig. 7.6 Contrast curve of the control moment exerted by the actuator

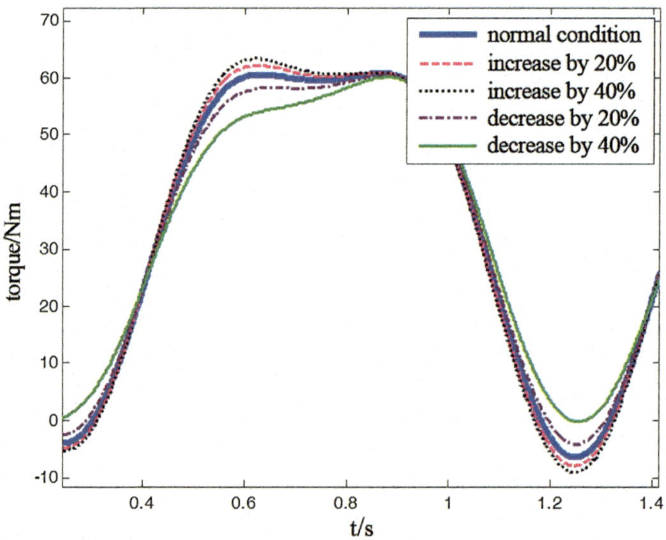

Fig. 7.7 Partial enlarged drawing of Fig. 7.6

regardless of whether there is any uncertainty, except the start phase, the neural network can estimate the model uncertainty existed in system model well. Therefore, it makes compensation, thereby reducing the energy consumption of the human body, and the desired design is achieved.

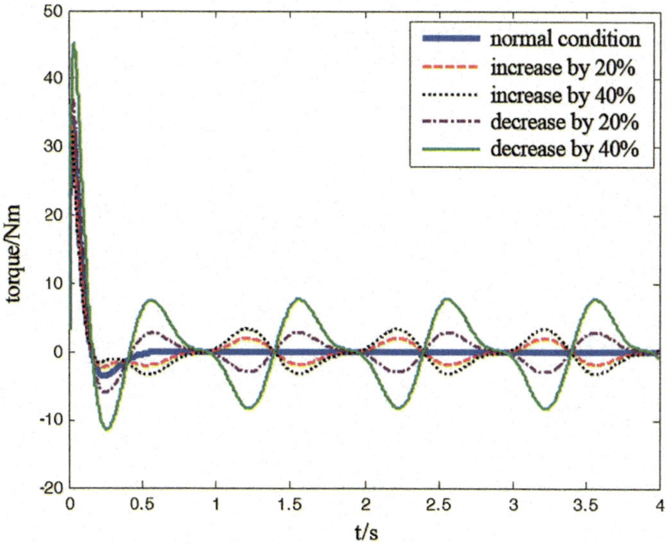

Fig. 7.8 Contrast curve of the human-machine acting force

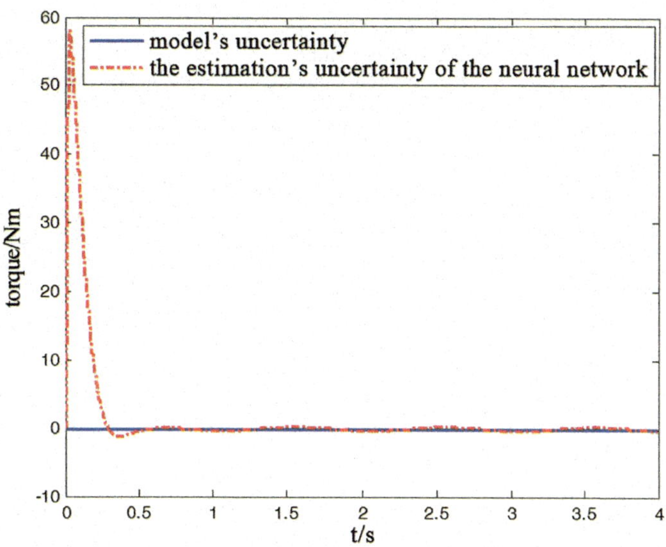

Fig. 7.9 Estimation of the neural network for the model uncertainty (without certainty)

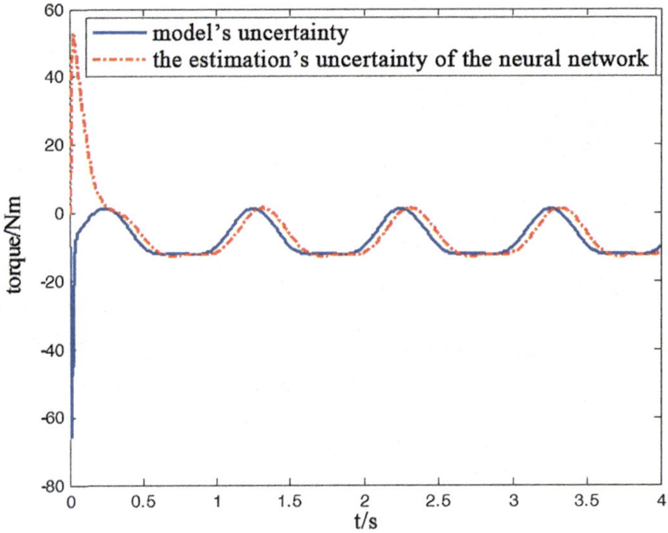

Fig. 7.10 Estimation of the neural network for the model uncertainty (the uncertainty increases by 20%)

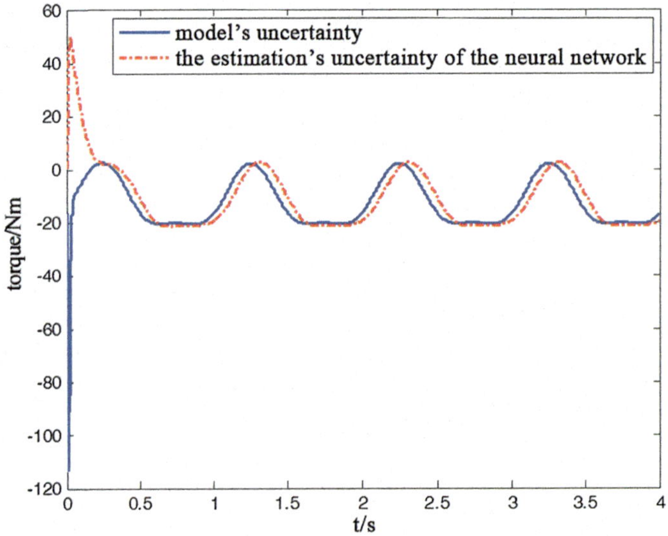

Fig. 7.11 Estimation of the neural network for the model uncertainty (the uncertainty increases by 40%)

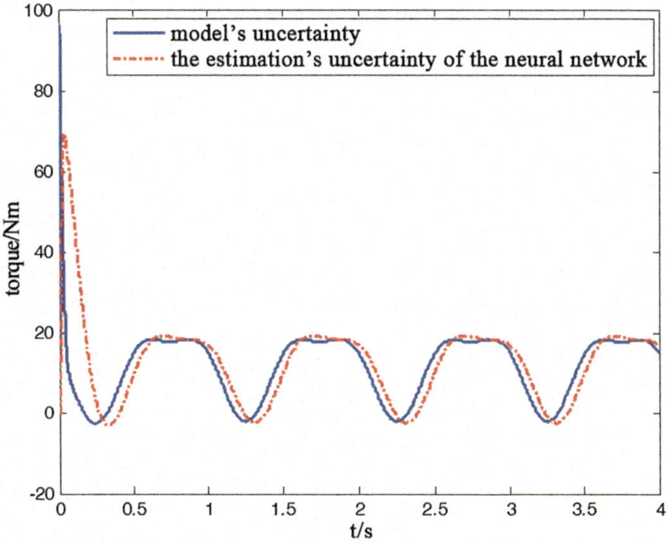

Fig. 7.12 Estimation of the neural network for the model uncertainty (the uncertainty decreases by 20%)

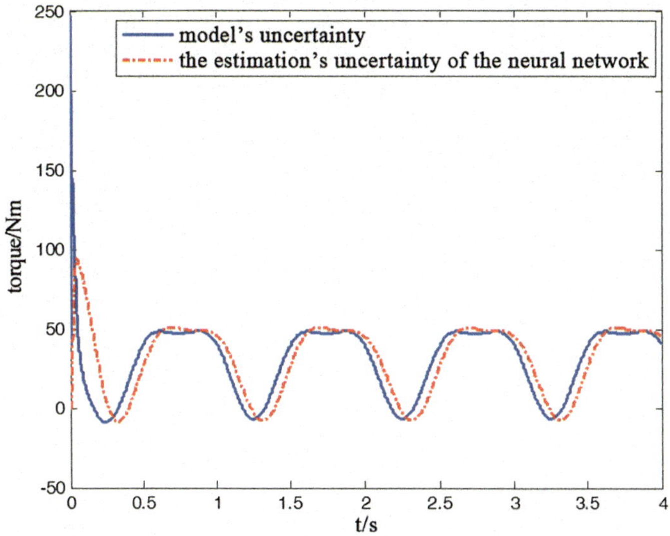

Fig. 7.13 Estimation of the neural network for the model uncertainty (the uncertainty decreases by 40%)

References

1. Wen Shuhuan. Intelligent Control and Trajectory Tracking Experiment Research of Uncertain Robot Force Position [D]. Yanshan University, Doctoral Dissertation, Qin Huangdao: 2004.
2. Jin Yuqiang. Robust Adaptive Control Research of Block Control Uncertain Nonlinear System [D]. Naval Aeronautical Engineering Institute, Doctoral Dissertation, Yantai: 2006.
3. Chang Xiaoguang, Li Wei, Fred. Wall. Experiment Study of a Mechanical Manipulator Under Uncertainty by Hybrid Fuzzy P+ID Controllers [J]. Control Theory and Applications, 2000, 17 (1): 73–78.
4. Chen Zengqiang, Lu Zhao, Yuan Zhuzhi. Neural Network's Variable Structure Control of the Uncertain System [J]. ActaScientiarum Naturallum Universitatis Nankaiensis, 2001, 34 (4): 48–52.
5. Hu Jianbo, Chu Jian. Variable Structure Control of a Class of Systems with Mismatched Uncertainties [J]. Control Theory and Applications, 2002, 19 (1): 105–108.
6. Luo Weilin, Lin Tong. RLED Robust Control with Uncertainty Based on Neural Network [J]. Journal of Fuzhou university, 2002, 30 (6): 822–826.
7. William, L. Oberkampf, Sharon M. Deland, Brain M. Rutherford, Kathleen V. Diegert, Kenneth F. Alvin. Error and Uncertainty in Modeling and Simulation [J]. Reliability Engineering and System Safety, 2002, 75 (1): 333–357.
8. Young-Hoon, Roh, Jun-Ho Oh. Sliding mode control with uncertainty adaptation for uncertain input-delay systems [J]. International Journal of Control, 2000, 73 (13): 1255–1260.
9. Santosh, Devasia. Robust Inversion-based Feedforward Controllers for Output Tracking under Plant Uncertainty [C]. Proceedings of the American Control Conference, Chicago, Illinois, USA, 2000: 497–502.
10. Wang Enping, Huanglin, Geng Zhiyong. Robust Control of the System with a Class of Uncertain Parameter [J]. Acta Automatica Sinica, 2001, 27 (1): 70–74.
11. Liu Jinkun. The Design and Matlab Simulation of Robot Control System [M]. Beijing: Tsinghua University Press, 2008.

Chapter 8
Exoskeleton Suit's Reference Trajectory Estimated Method Based on Neural Network

8.1 Introduction

In Chap. 6, we use human–machine interaction model to estimate the reference trajectory, reference speed, and reference acceleration of the exoskeleton suit, but the precise form of human–machine acting force model cannot be known in advance. Using the fixed human–machine acting force model cannot estimate the intention of the human accurately (i.e., the reference trajectory of the exoskeleton suit); for example, in different motion states, the forms and parameters of the human–machine acting force are different (linear or nonlinear or order). While the use of the neural network can find the nonlinear mapping between the human–machine acting force and human's movement trajectory, the movement reference trajectory of the exoskeleton suit is estimated correctly. Therefore, in this chapter, we put forward the method of using neural network to estimate the reference trajectory of the exoskeleton suit, whose principle is as shown in Fig. 8.1.

8.2 Reference Trajectory Estimated Method Based on Neural Network

Radial basis function (RBF) neural network is a kind of neural network put forward by J. Moody and C. Darken in the late 1980s, which is with a three-layer feed-forward network of single hidden layer. RBF network simulates the neural network structure of local adjustment in human brain, mutual coverage accepted domain (or receptive field), to prove that the RBF network can approximate arbitrary continuous function in arbitrary precision.

RBF network learning process is similar to the learning process of BP network, and their main difference is that they use different action functions. In BP network, the hidden layer uses sigmoid function, and its value is nonzero within the infinite

Z. Yang et al., *Force Control Theory and Method of Human Load Carrying Exoskeleton Suit*, DOI 10.1007/978-3-662-54144-9_8

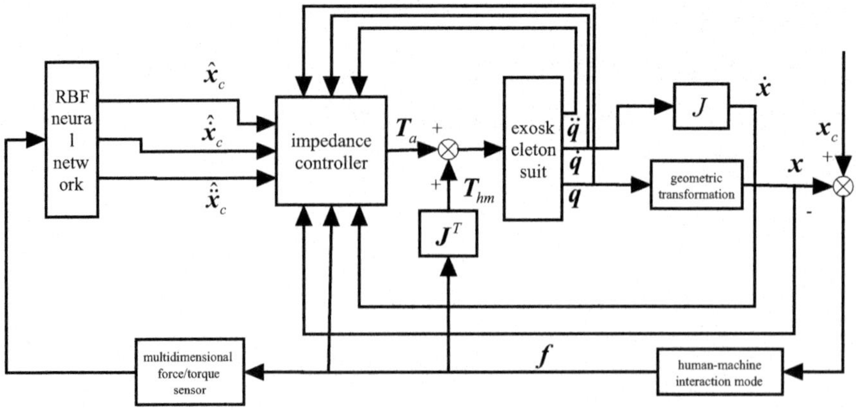

Fig. 8.1 Exoskeleton suit's reference trajectory estimated method based on neural networks and its application in the impedance control

range in input space, so it is a kind of neural network of global approximation. While the action function in RBF network function is gaussian basis function, its value within the limited range in input space is nonzero, so the RBF network is the neural network of local approximation.

RBF network is a kind of three-layer forward network. The mapping from input to output is nonlinear, and the mapping from hidden layer space to output space is linear. The RBF network is the neural network of local approximation, so the use of RBF network can significantly accelerate the learning speed and avoid local minimum problem, which is suitable for the requirement of real-time control. Adopting RBF network to constitute the neural network control scheme can effectively improve the accuracy, the robustness, and self-adaptability of the system.

RBF network identification structure is shown in Fig. 8.2.

RBF network's radial base vector is $\boldsymbol{H} = [h_1, h_2, \ldots, h_m]^{\mathrm{T}}$, in which h_i is the gaussian basis function [1]

$$h_i = f(\|z - \varphi_i\|/\sigma_i^2), \quad i = 1, 2, \ldots, m \tag{8.1}$$

Fig. 8.2 RBF neural network structure

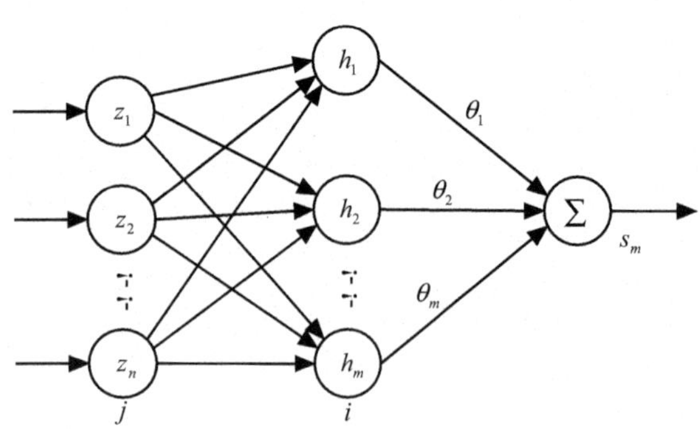

where $z = [z_1, z_2, \ldots, z_n]^{\mathrm{T}}$ is the network's input signal; $\boldsymbol{h}(z) = [h_1, h_2, \ldots, h_m]^{\mathrm{T}}$ is the output of the gaussian basis function; m represents the number of neurons; and $\varphi_i = [\varphi_{i1}, \varphi_{i2}, \ldots, \varphi_{in}]$ represents the center vector of the network's number i node.

We set the network's base width vector as

$$\boldsymbol{\sigma} = [\sigma_1, \sigma_2, \ldots, \sigma_m]^{\mathrm{T}} \tag{8.2}$$

σ_i is the base width parameter of node i and is the number greater than zero.

Network weights vector is as follows:

$$\boldsymbol{\theta} = [\theta_1, \theta_2, \ldots, \theta_m]^{\mathrm{T}} \tag{8.3}$$

The output of the RBF network is as follows:

$$s_m(t) = \boldsymbol{\theta}^{\mathrm{T}} \boldsymbol{h}(z) \tag{8.4}$$

The performance index function of RBF network is as follows:

$$J = \frac{1}{2}\left(s_m(t) - s(t) - \frac{f}{\hat{K}_{\mathrm{pf}}}\right)^2 \tag{8.5}$$

where $\hat{K}_{\mathrm{pf}}$ is the estimated rigidity coefficient of human–machine interaction model (here it is the one-dimensional case of the stiffness coefficient matrix, and $\hat{K}_{\mathrm{Pf}}$ represents the estimated value for K_{Pf}); f is the human–machine interaction force obtain by measurement. According to the gradient descent method, the iterative algorithm of the output weights is as follows:

$$\theta_i(t) = \theta_i(t-1) + \eta\left(s_m(t) - s(t) - \frac{f}{\hat{K}_{\mathrm{pf}}}\right)h_i + \alpha(\theta_i(t-1) - \theta_i(t-2)) \tag{8.6}$$

Parameters of node center and node width are unchanged.

If using the neural network and using the human–machine interaction force as the neural network input signals, we need not the training sample when we estimate the real-time reference trajectory of the exoskeleton suit online.

8.3 Simulation

For the sake of simplicity, we select the controlled object to be the single degree of freedom model and simulate in the M language environment of MATLAB. The exoskeleton suit of single degree of freedom is considered to be a single pendulum that is tied together with a person, and it is driven by hip joint actuator and the human–machine acting force. During simulation, it defines the mass attributes and length of the exoskeleton suit to be 10 kg and 0.5 m, respectively. For the human–machine acting force model, we adopt the linear combination of rigidity, spring, and damping [2, 3], as shown in Eq. (8.7).

$$f = \boldsymbol{K}_{\mathrm{Pf}}(\boldsymbol{x}_{\mathrm{h}} - \boldsymbol{x}_{\mathrm{e}}) + \boldsymbol{K}_{\mathrm{Df}}(\dot{\boldsymbol{x}}_{\mathrm{h}} - \dot{\boldsymbol{x}}_{\mathrm{e}}) + \boldsymbol{K}_{\mathrm{Mf}}(\ddot{\boldsymbol{x}}_{\mathrm{h}} - \ddot{\boldsymbol{x}}_{\mathrm{e}}) \tag{8.7}$$

where f represents the acting force of human exerting on the exoskeleton suit; $\boldsymbol{x}_{\mathrm{h}}, \dot{\boldsymbol{x}}_{\mathrm{h}}, \ddot{\boldsymbol{x}}_{\mathrm{h}}$ represent the movement information of human (the position, speed, and acceleration in operation space); $\boldsymbol{K}_{\mathrm{Pf}}$, $\boldsymbol{K}_{\mathrm{Df}}$, and $\boldsymbol{K}_{\mathrm{Mf}}$ represent the spring coefficient (rigidity coefficient) matrix, damping coefficient matrix, and inertia coefficient matrix, respectively.

Without considering the inertia coefficient, Eq. (8.7) also can be simplified to the spring damping model

$$f = \boldsymbol{K}_{\mathrm{Pf}}(\boldsymbol{x}_{\mathrm{h}} - \boldsymbol{x}_{\mathrm{e}}) + \boldsymbol{K}_{\mathrm{Df}}(\dot{\boldsymbol{x}}_{\mathrm{h}} - \dot{\boldsymbol{x}}_{\mathrm{e}}) \tag{8.8}$$

If we consider without inertia coefficient and damping coefficient, Eq. (2.106) also can be simplified to the spring model [4]

$$f = \boldsymbol{K}_{\mathrm{Pf}}(\boldsymbol{x}_{\mathrm{h}} - \boldsymbol{x}_{\mathrm{e}}) \tag{8.9}$$

The acting force of human exerting on the exoskeleton suit and the acting force of the exoskeleton suit exerting on human are the reactive forces for each other, and both can be called human–machine acting forces. In this book, the human–machine force refers to the acting force of human exerting on exoskeleton suit.

During simulation, we have designed the following experiments under the six kinds of cases, and they are as follows:

(1) Simulation without adding control

In this simulation, the exoskeleton suit's external actuator $\boldsymbol{T}_{\mathrm{a}} = 0$, through the human–machine interaction model, human exerts human–machine applied moment $\boldsymbol{T}_{\mathrm{hm}}$ on the exoskeleton suit; the simulation includes three experiments, and different human–machine interaction model parameters are selected.

$K_{\mathrm{Pf}} = 500,\ K_{\mathrm{Df}} = 0,\ K_{\mathrm{Mf}} = 0$

(1) experiment 1: $K_{\mathrm{Pf}} = 500,\ K_{\mathrm{Df}} = 10,\ K_{\mathrm{Mf}} = 0$

$K_{\mathrm{Pf}} = 500,\ K_{\mathrm{Df}} = 8,\ K_{\mathrm{Mf}} = 0.005$

(2) experiment 2: $K_{\mathrm{Pf}} = 500,\ K_{\mathrm{Df}} = 10,\ K_{\mathrm{Mf}} = 0$

(3) experiment 3: $K_{\mathrm{Pf}} = 500,\ K_{\mathrm{Df}} = 8,\ K_{\mathrm{Mf}} = 0.005$

(2) Simulation with control

In this simulation, the exoskeleton suit' movement is caused by the joint function of control moment $\boldsymbol{T}_{\mathrm{a}}$ and human–machine interaction torque $\boldsymbol{T}_{\mathrm{hm}}$; the control moment $\boldsymbol{T}_{\mathrm{a}}$ is determined by Eq. (6.10); and the reference trajectory is identified by neural network. We set the neural network parameters as follows: The number of neurons is 10; basis function is gaussian basis function; the node center parameter is

0.8; the node base width parameter is 20; the network learning rate is $\eta = 0.9$; momentum factor is $\alpha = 0.9$. The desired impedance coefficient, respectively, is as follows: $K_{\mathrm{Pp}} = 1, K_{\mathrm{Dp}} = 3, K_{\mathrm{Mp}} = 0.01$. The estimated human–machine interaction model stiffness parameter is $\hat{K}_{\mathrm{Pf}} = 600$. In order to contrast the control effect under different human–machine interaction model parameters, the simulation also is divided into three experiments, and different human–machine interaction model parameters are selected.

(1) experiment 4: $K_{\mathrm{Pf}} = 500,\ K_{\mathrm{Df}} = 0,\ K_{\mathrm{Mf}} = 0$
(2) experiment 5: $K_{\mathrm{Pf}} = 500,\ K_{\mathrm{Df}} = 10,\ K_{\mathrm{Mf}} = 0$
(3) experiment 6: $K_{\mathrm{Pf}} = 500,\ K_{\mathrm{Df}} = 2,\ K_{\mathrm{Mf}} = 0.006$

8.3.1 Simulation Without Control

The simulation results of the experiment 1 are as shown in Figs. 8.3 and 8.4. Figure 8.3 shows the joint angle tracking curve of the exoskeleton suit, and Fig. 8.4 shows the torque curve exerted by human, which illustrates that under the function of simple rigid human–machine interaction model, the exoskeleton suit's angle tracking effect for the human body is relatively poor, and the torque needs to be exerted by human is also large.

The simulation results of the experiment 2 are shown in Figs. 8.5 and 8.6. From the Fig. 8.6, we can see that the use of the human–machine interaction model of spring damping kind can achieve the skeleton suit's good tracking for human motion, and the torque exerted by human is also smaller.

The simulation results of the experiment 3 are shown in Figs. 8.7 and 8.8. And its simulation results are similar to those of the experiment 2.

The three experiments show that parameters of the human–machine interaction model have effect on simulation results; its physical meaning is that when different

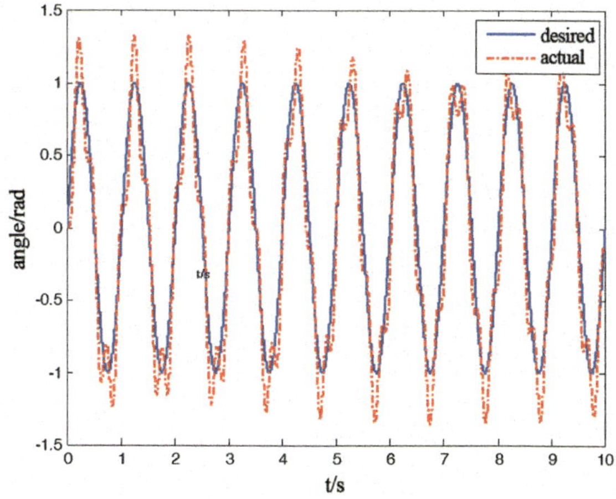

Fig. 8.3 Angle tracking curve (experiment 1)

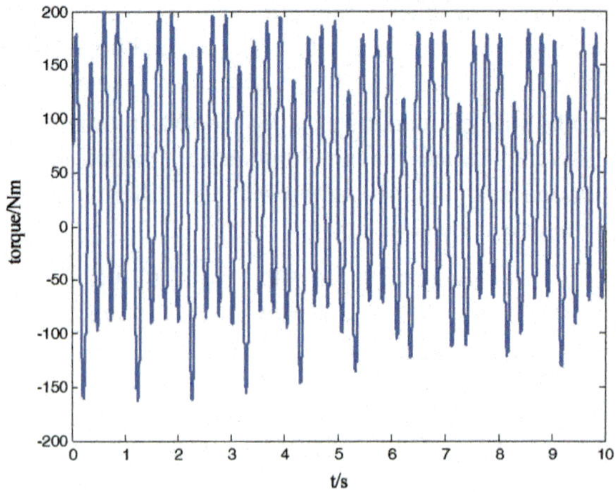

Fig. 8.4 Torque exerted by human (experiment 1)

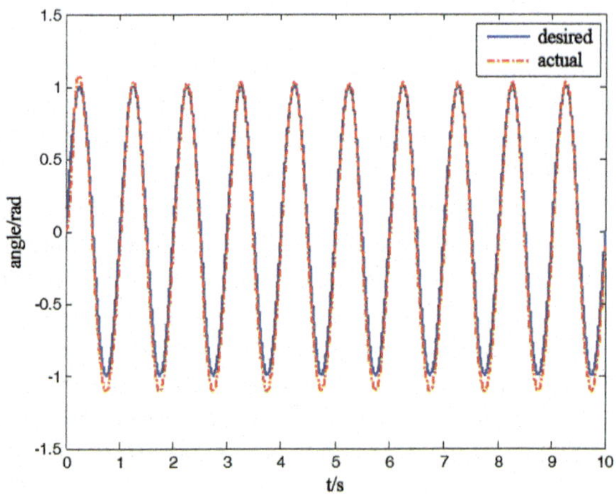

Fig. 8.5 Angle tracking curve (experiment 2)

people swing the exoskeleton suit of the single degree of freedom, the obtained tracking curve is different, and its essence is that the human–machine interaction model parameters are different. This difference causes that if we use the fixed human–machine interaction model parameter to estimate the system's reference trajectory, the estimation is bound to be not accurate, thus causing the human–machine force to increase and then increasing the human body's energy consumption.

8.3.2 Simulation with Control

When the control is added, the exoskeleton suit's movement is driven by the joint function of the exoskeleton suit itself and the human body, and our goal is through

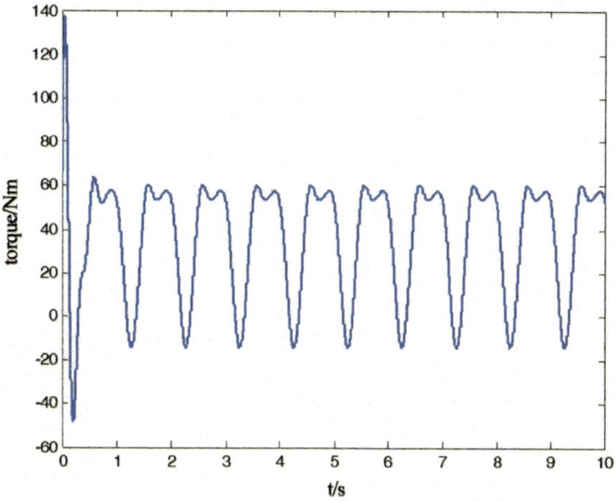

Fig. 8.6 Torque exerted by human (experiment 2)

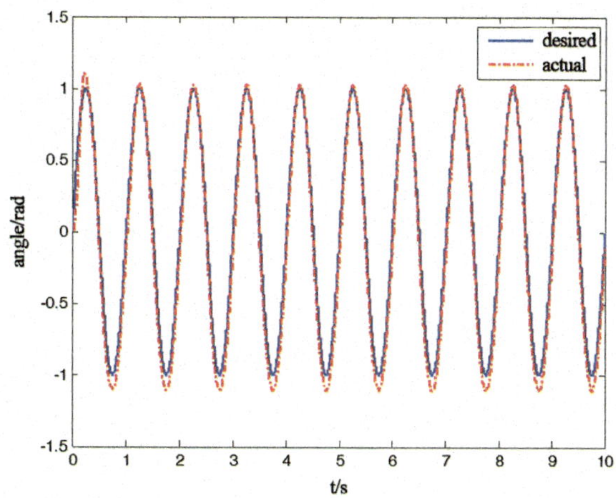

Fig. 8.7 Angle tracking curve (experiment 3)

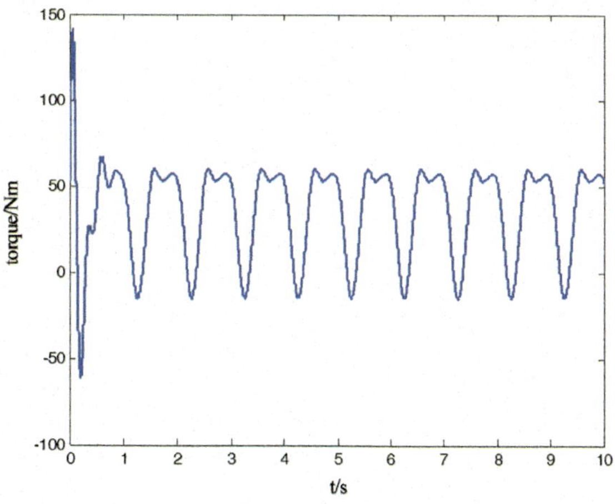

Fig. 8.8 Torque exerted by human (experiment 3)

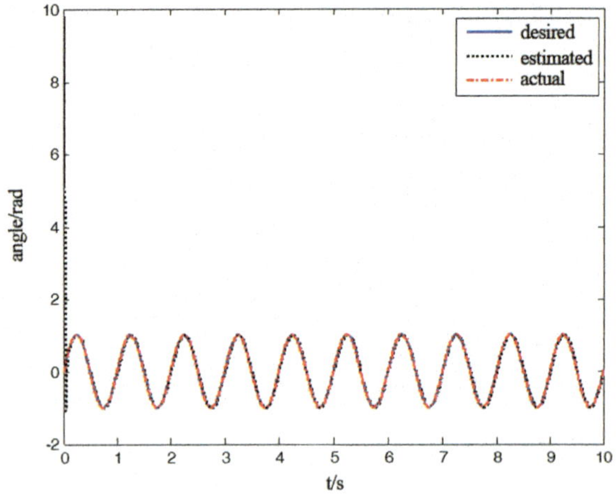

Fig. 8.9 Angle tracking curve (experiment 4)

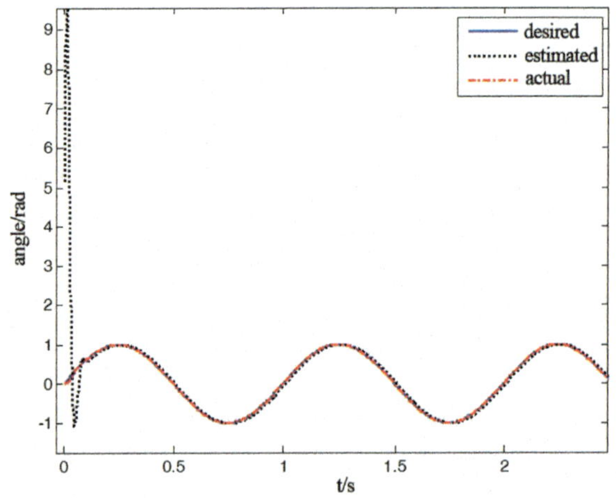

Fig. 8.10 Local amplification of Fig. 8.9

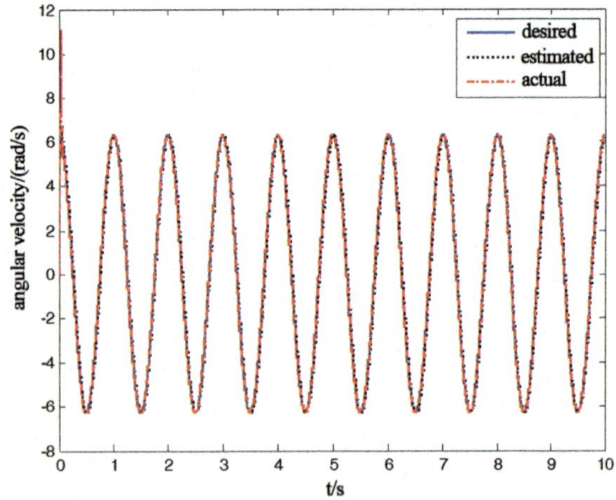

Fig. 8.11 Angle tracking curve (experiment 4)

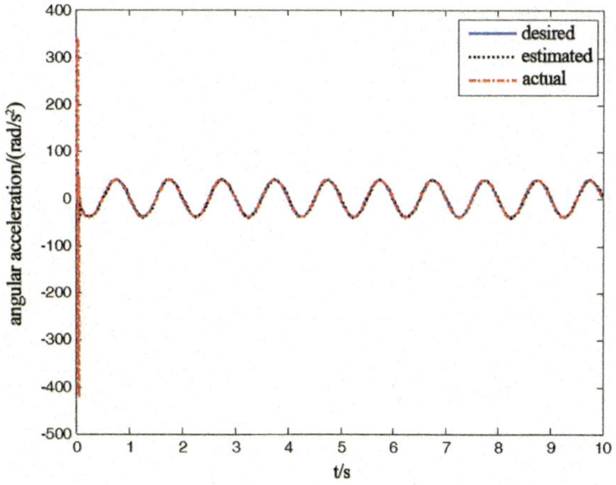

Fig. 8.12 Angle acceleration tracking curve (experiment 4)

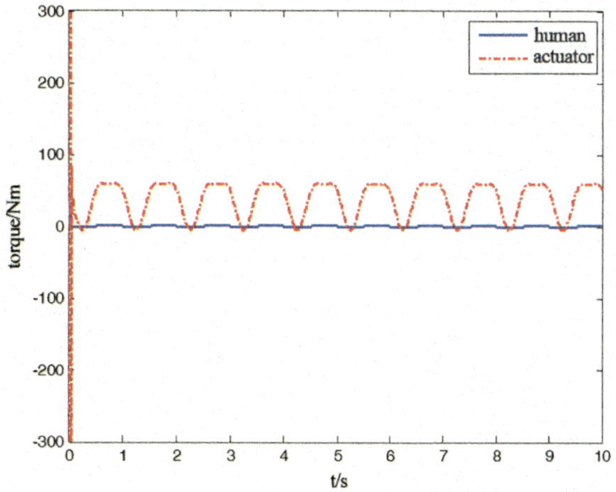

Fig. 8.13 Contrast curve between the torque exerted by the actuator and the torque exerted by human (experiment 4)

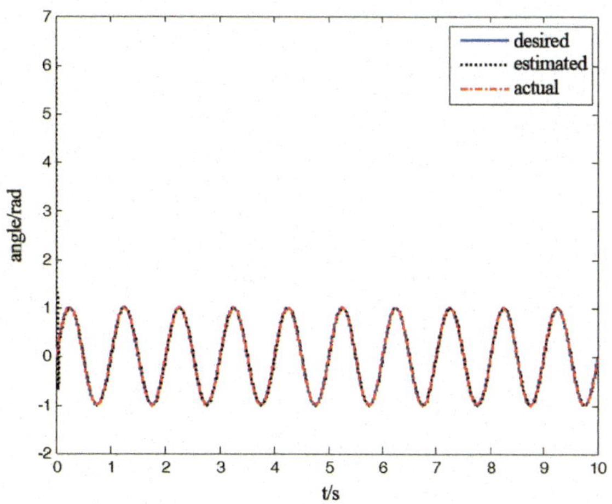

Fig. 8.14 Angle tracking curve (experiment 5)

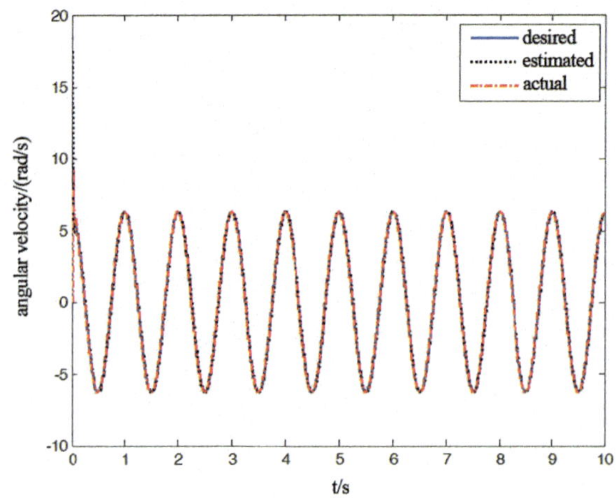

Fig. 8.15 Angular velocity tracking curve (experiment 5)

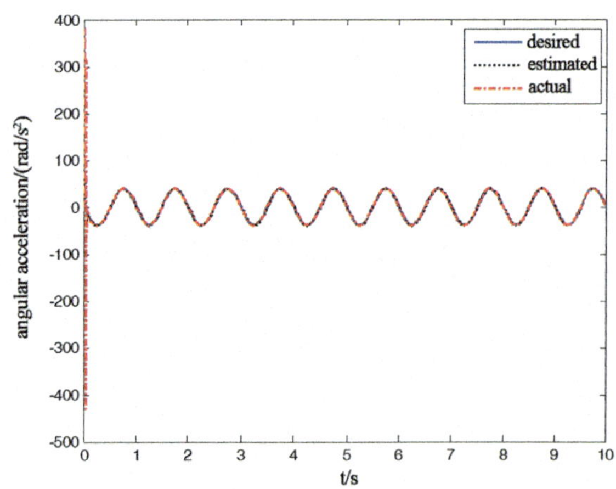

Fig. 8.16 Angular acceleration tracking curve (experiment 5)

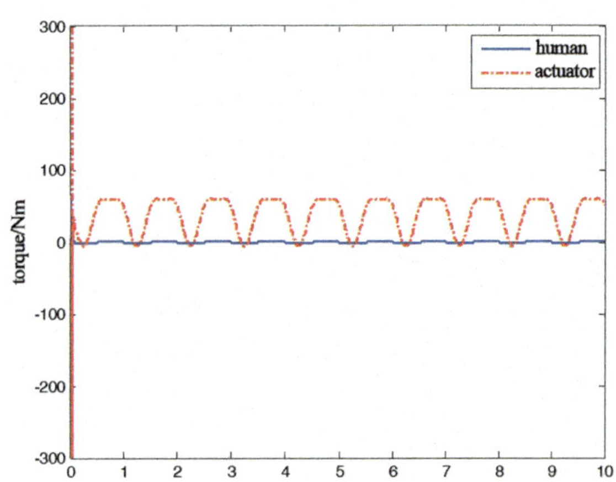

Fig. 8.17 Contrast curve between the torque exerted by the actuator and the torque exerted by human (experiment 5)

the human–machine acting force to estimate the reference trajectory of the exoskeleton suit and to generate the driving moment of the exoskeleton suit itself with the impedance controller, thus reducing the torque exerted by the human body and then reducing the energy consumption of the human body.

Figures 8.9, 8.10, 8.11, 8.12, and 8.13 show the simulation results of experiment 4. Among them, Fig. 8.9 gives the angle tracking curve, and Fig. 8.10 is the local amplification of Fig. 8.9. From the two figures, we can see that the desired angle signal estimated through neural network is very accurate, and after the joint function of human body and the controller, the exoskeleton suit's tracking angle for the human body is also very accurate.

Figure 8.11 shows the contrast curve among the desired reference angular velocity signal, the estimated reference angular velocity signal, and the exoskeleton

Fig. 8.18 Angle tracking curve (experiment 6)

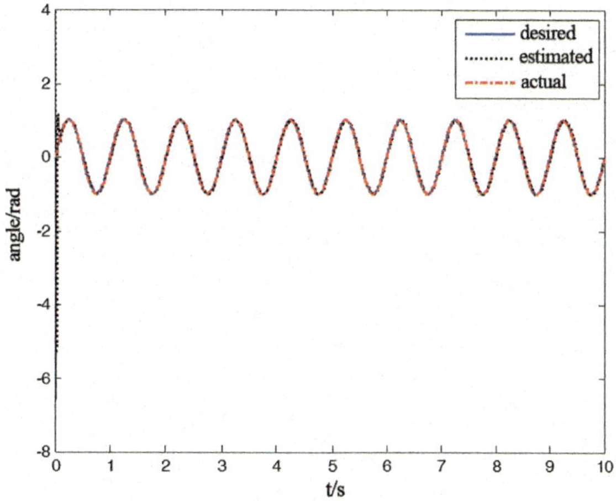

Fig. 8.19 Angular velocity tracking curve (experiment 6)

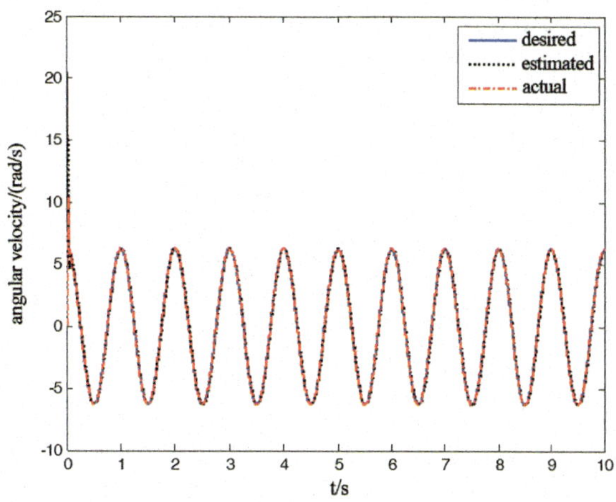

suit's actual angular velocity signal. Figure 8.12 shows the contrast curve among the desired reference angular acceleration signal, the estimated reference angular acceleration signal, and the actual angular acceleration signal. Figure 8.13 shows the contrast curve between the control moment exerted by human and the control moment exerted by the actuator, which illustrates that in the process of the exoskeleton suit's movement, the actuator has exerted the main control moment, and the human body has exerted very small control moment, so the method can greatly reduce the energy consumption of the human body. Experiments 4, 5, and 6 all assume the estimated human–machine interaction model stiffness parameter $\hat{K}_{\mathrm{Pf}} = 600$, which has some difference from the actual human–machine interaction model parameter. In experiment 5, the spring damping model is used as the human–machine interaction model, and the simulation curve of the experiment 5 is shown in Figs. 8.14, 8.15, 8.16, and 8.17. Figures 8.14, 8.15, and 8.16 illustrate that the

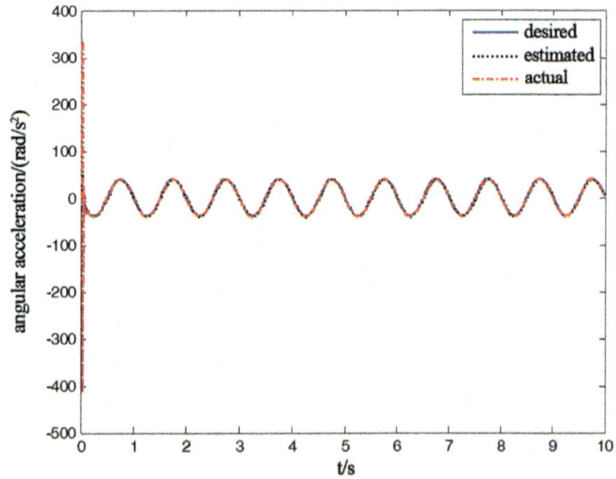

Fig. 8.20 Angular acceleration tracking curve (experiment 6)

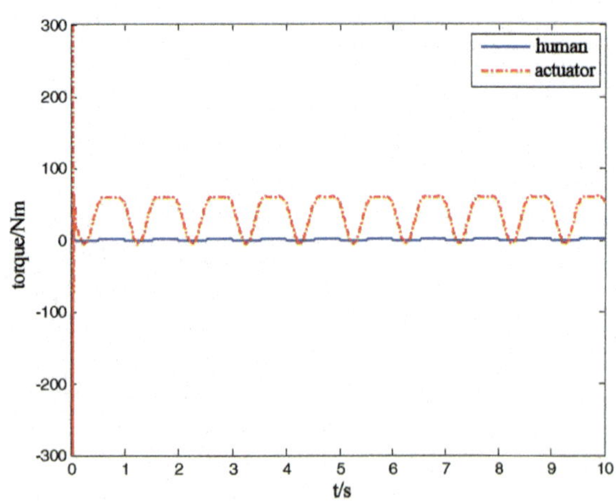

Fig. 8.21 Contrast curve between the torque exerted by the actuator and the torque exerted by human (experiment 6)

neural network's estimation for the reference trajectory is very accurate, and the exoskeleton suit's movement trajectory tracking for the reference trajectory is very accurate. Figure 8.17 illustrates that in the movement process of the exoskeleton suit, the control moment exerted by the exoskeleton suit's controller plays a major role, whereas the control moment exerted by the human body is very small (Figs. 8.18, 8.19, 8.20, and 8.21).

The results of the experiment 6 and experiment 5 are similar, except that actually use impedance model as the human–machine interaction model in the experiment 6. The three simulation experiments demonstrate that when there is a big difference between the estimated human–machine interaction model and the actual human–machine interaction model, RBF neural network can estimate the reference trajectory of the exoskeleton suit very well, thereby controlling the exoskeleton suit to follow the movement of the human body, greatly reducing the energy consumption of the human body, and eventually achieving the goal of the design.

References

1. Liu Jinkun. The Design and Matlab Simulation of Robot Control System [M]. Beijing: Tsinghua University Press, 2008.
2. Heng, Wang, K. H. Low, Michael Yu Wang. Reference Trajectory Generation for Force Tracking Impedance Control by Using Neural Network-based Environment Estimation [C]. 2006 IEEE Conference on Robotics Automation and Mechatronics, Bangkok, Thailand, 2006: 1–6.
3. Luo Yangyu, Wang Dangxiao, Zhang Yuru. Modeling and Analysis of a Force Interactive System [J]. Journal of Beijing University of Aeronautics and Astronautics, 2004, 30 (6): 539–542.
4. Murat, Cenk Cavusoglu, Joseph Yan, S. Shankar Sastry. A Hybrid System Approach to Contact Stability and Force Control in Robotic Manipulators [C]. In Proceedings of the 12th IEEE International Symposium on Intelligent Control (ISIC'97), Istanbul, Turkey, 1997: 143–148.

Chapter 9
Conclusions

9.1 Conclusions of Major Contents

Exoskeleton suit not only has important implication value for military but also can be widely used in civilian areas such as rescue, disaster relief, mountain climbing, construction, tourism, and helping the disabled. But it has been a difficult problem for the exoskeleton suit system getting to be mature how to make the exoskeleton suit perceive human body movement intentions and track human body movement while bearing load. This book has analyzed the differences between robot force control and exoskeleton suit force control and has made a deep study on the exoskeleton suit force control technology. Details are as follows.

(1) This book summarizes present status of exoskeleton suit technology at home and abroad and focuses on the investment and technical research results of exoskeleton suit in the USA. It gives an overview of the control methods which have been applied in exoskeleton suit system and analyzes their advantages and disadvantages. It introduces the dynamics and kinematics model of exoskeleton suit combined with a deep analysis of the human-machine interaction model that lays a foundation for the research of the exoskeleton suit.

(2) This book presents a sensitivity amplification control method of exoskeleton suit. It elaborates the fundamental thought and theory basis of sensitivity amplification control method. And theoretically, the stability of the method has been proved with the nonlinear model of exoskeleton suit. The exoskeleton suit sensitivity amplification control simulation of human-machine integration has been realized in the computer adopting classical human-machine interaction model, and the comparison results are given. For the disadvantage that the sensitivity amplification control method needs precise mathematical model of the object, it proposes a sensitivity amplification control method of exoskeleton suit based on the neural network; the method's stability has been proved, and the simulation results have been given.

Z. Yang et al., *Force Control Theory and Method of Human Load Carrying Exoskeleton Suit*, DOI 10.1007/978-3-662-54144-9_9

(3) This book analyzes similarities and differences of force control methods between the traditional robot and the exoskeleton suit. It views trunk center and ankle of exoskeleton suit as two end effectors in different models and proposes a force control method of exoskeleton suit that installing the multi-dimensional force/torque sensors on the end effectors as force feedback, whose concept is used as the ideological basis of Chaps. 4, 5, 6, and 7. Meanwhile, four kinds of force control methods of the exoskeleton suit are introduced and a lot of simulation studies are made, as follows:

   (1) Direct force control method of exoskeleton suit: using the static model compensation method, it generates the control moment of joint space directly after making the feedback force Jacobian conversion, to control the movement of exoskeleton suit and ensure human-machine acting force smaller; in other words, this method reduces the control force of human exerting on the exoskeleton suit, thereby reducing the human body's energy consumption. In this book, controller's stability has been proved in three kinds of human-machine interaction models, and the controller's effectiveness and correctness have been proved by the given simulation results.
   (2) The force control method of exoskeleton suit based on the inner position loop: first of all, this method is to design position control loop in the exoskeleton suit to ensure accurate movement of the exoskeleton suit according to the reference trajectory. And then through force feedback information of the multi-dimensional force/torque sensors, the reference trajectory of inner position loop in operation space is generated so as to control the exoskeleton suit to follow human body's movement. We design two kinds of inner position loop controllers based on the static model and dynamic model and make simulation study for them separately. The simulation results prove the force control method based on the inner position loop can reduce the human-machine interaction force effectively and have strong robustness for the changes of the operator and load.
   (3) Impedance control method of exoskeleton suit: Impedance control method is one of the classical force control methods. Impedance control depends on adjusting the dynamic relation between the position of end effectors and contact force to realize a compliance control with the desired impedance characteristics. However, impedance control usually needs to use force information to estimate the reference trajectory correction, and the reference trajectory, reference velocity, reference accelerated velocity of the system are known. The proposed impedance control of exoskeleton suit is different from the general robot impedance control, and its reference trajectory, reference velocity, and reference accelerated velocity are unknown. So it needs to estimate not only the reference trajectory but also the reference velocity and reference accelerated velocity. Generally speaking, we only need to estimate reference trajectory, and we can get the signals of reference velocity and reference accelerated velocity by

difference methods. But because of the noise signal in the force feedback information, difference method may make the signals of reference velocity and reference accelerated velocity unavailable. Therefore, we propose an impedance control method of exoskeleton suit based on the tracking differentiator which has solved the problem of estimating reference velocity and reference accelerated velocity in high-noise condition.

(4) Impedance control method of exoskeleton suit with uncertainty: This system model is built for a standard model when designing impedance controller of exoskeleton suit. But in reality, the load carried by exoskeleton suit often needs changing with a wide range. So the controller designed according to a standard model cannot meet the performance requirements of the load changes, and human-machine acting force will increase, thus causing the increase of energy expenditure. Concerning this issue, we propose an impedance control method of exoskeleton suit adopting RBF neural network to estimate system's uncertainty, and its stability has been proved. The simulation results show that the neural network is more accurate to estimate system's uncertainty, thus reducing the influence of system's uncertainty to human-machine acting force significantly.

(4) This book introduces the reference trajectory estimated method of exoskeleton suit based on the neural network. Generally, if the human-machine acting force model is known, then the model can be used to estimate the human body's movement trail, namely the exoskeleton suit's reference trajectory. However, because of the uncertainty of the human-machine interaction model parameter, there are large errors in estimated reference trajectory which will cause the increase of human-machine interaction force. As a solution to this problem, we suggest an adaptive method of estimating exoskeleton suit reference trajectory using RBF neural network. This method can estimate the exoskeleton suit reference trajectory in real-time online and has strong robustness for the change of human-machine interaction model parameter.

9.2 Comparing and Analysis of the Control Method

From this book, we can see that the force control method has many advantages compared with the sensitivity amplification control method. That is to say, the increase of force feedback information can make the controller be aware of its control effect, so it is easy to reduce the human-machine acting force further by using intelligent control algorithm. So the force control method is superior to the sensitivity control method.

It is hard to say which method is better intuitively for these force control methods we studied all have their own advantages and disadvantages. In order to compare and analyze their advantages and disadvantages, we designed a mark sheet of control methods comparison as shown below. In this process, a score between 1 and 5 is given to every control performance of the studied item according to its relative merits. The item with best performance gets the highest score, and the final score of each control method is obtained by adding the score of each studied item. So we can judge the relative merits of control methods according to the score scale.

Items studied in Table 9.1 are described as follows:

(1) Controller structure: refers to the complexity of control algorithm and the magnitude of calculated amount. Direct force control method and the force control method based on the inner position loop have less calculated amount for adopting static model to design controller, thus getting higher score. But impedance control method and the impedance control method with uncertainty have complex algorithm and large amount of calculation because the controller is designed based on the dynamic model of exoskeleton suit, thus getting lower score.
(2) Tracking error: refers to the magnitude of trajectory tracking error. Because this item is very important for measuring the controller's performances, it takes a large proportion. Therefore, its score is multiplied by 2.
(3) Dynamic performance: refers to the vibration condition of human-machine acting force in control process.
(4) Human–machine acting force: refers to the magnitude of human-machine acting force in control process. Taking two aspects of human factor and maximum human factor into account synthetically, this item is the most important index for measuring the force controller. So its score is also doubled.
(5) Robustness for the change of human-machine interaction model parameter: refers to the increase and decrease conditions of human-machine acting force when the human-machine interaction model parameter changes.
(6) Robustness for load variation: refers to the increase and decrease conditions of human-machine acting force when the load changes.

The composite score shows that the impedance control method with uncertainty has the highest score, but it is not quite different from the impedance control method. Although the structure of direct force control method is the simplest, its other items' scores are lower, so its composite score is the lowest. However, the comparison method shown by Table 9.1 only has a guiding role. In reality, more examined items are needed and they should be assigned corresponding weights when the control method is selected.

9.3 Outlook

The research on the exoskeleton suit is an international leading-edge project covering many subjects. It is a quite difficult project to estimate body movement intentions conveniently and accurately in real time. This book adopts

Table 9.1 Mark sheet of control methods comparison

Examined items	Direct force control method	The force control method based on the inner position loop	Impedance control method	The impedance control method with uncertainty
Controller structure	5	4	2	1
Tracking error	6	2	10	10
Dynamic performance	2	5	4	4
human-machine acting force	2	6	10	10
Robustness for the change of human-machine interaction model parameter	3	5	4	4
Robustness for load variation	1	5	2	4
Total points	19	27	32	33

multi-dimensional force/torque sensors to measure human-machine acting force for judging the human body movement intentions, and we put forward a series of force control methods of exoskeleton suit. But due to the limitation of time and experimental conditions, a lot of work needs to be done further, such as follows:

(1) Because of the limited fund, multi-dimensional force/torque sensors are expensive, so we cannot help saying that it is a pity without experimental verification for the designed force control methods. We believe that there must be more undesired discoveries by the oncoming experimental study.
(2) Due to the difficulty of simulating human walking in a computer, we have only studied the force control of exoskeleton suit in support state. So we need to research on that of all-direction gait later in order to get more comprehensive performances of force control.
(3) Because there is no experience in applying force control to exoskeleton suit at home and abroad, what has been done described in this book is based on the classical force control method. Advanced force control strategy can be brought in the exoskeleton suit control further in the following study, such as adaptive control technology, robust force control, learning force control, neural network force control, and model predictive force control.

Appendix A
Calculate Method of Human Attribute

```
%-----------------------------------
%                 humandata.m
%-----------------------------------
% - Units: kg,m,N,N.m.
%     -     Anthropometric     data     computed     from     Winter     D.
A., Biomechanics of Human Movement, 1979
% and Andersson, Occupational Biomechanics (1991)

H = 1.75;                  % Human height
M = 70;                   % Total mass of human

mf = 0.0143*M;               % Mass of foot
Lf = 0.152*H-0.1             % Length of foot along sole without toe
LGf = Lf-0.5*0.152*H           % distance from toe joint to foot CG
hGf = 0;
If = mf*(0.475*Lf)^2;          % Inertia of foot about its CG
hf = 0.039*H             % Distance from ankle to foot sole

ms = 0.0465*M;               % Mass of shank
Ls = 0.285*H;               % Length of shank
LGs = 0.567*Ls;              % Distance from ankle to shank CG
hGs = 0;
Is = ms*(0.302*Ls)^2;          % Inertia of shank about its CG

mt = 0.1*M;                 % Mass of thigh
Lt = 0.245*H;               % Length of thigh
LGt = 0.567*Lt;              % Distance from knee to thigh CG
hGt = 0;
It = mt*(0.323*Lt)^2;          % Inertia of thigh about its CG
```


Z. Yang et al., *Force Control Theory and Method of Human Load Carrying Exoskeleton Suit*, DOI 10.1007/978-3-662-54144-9

```
mub = M-2*mt-2*ms-2*mf;        % Mass of upper body
Lub = 0.288*H;                 % Length of upper body (hip to shoulder)
LGub = 0.374*Lub;              % Vertical Distance from hip to upper body CG
hGub = -0.05;  % Horizontal Distance from hip to upper body CG (+ve =forward)
Iub = mub*(0.496*Lub)^2;       % Inertia of upper body about its CG

LI = Ls + Lt;                  % Leg length
mI = ms + mt;                  % Mass of leg
LGI = (ms*LGs+mt*(Ls+LGt))/(ms+mt);        % Distance of leg CG from ankle
II     =     Is     +     ms*(LGI-Ls)^2+It+mt*(LGI-(Ls+LGt)
^2);   % Inertia of leg about its CG
g = 9.81;                      % Gravitational constant (m/s2)
```

Appendix B
Clinical Gait Analysis Data

Ankle joint angle	Knee joint angle	Hip joint angle	Trunk angle	Ankle joint torque	Knee joint torque	Hip joint torque	Ankle joint energy	Knee joint energy	Hip joint energy
2.659	6.464	28.813	9.463	0	−0.202	0.299	0	0.018	0.111
1.432	7.102	28.237	9.449	0.021	−0.365	0.715	0.017	0.319	0.356
0.27	8.449	27.53	9.359	0.023	−0.136	0.41	0.011	0.149	0.229
−0.328	10.158	26.662	9.188	0.071	−0.034	0.326	0.004	0.011	0.225
−0.104	11.772	25.546	8.958	0.128	0.058	0.247	−0.069	−0.116	0.214
0.951	12.992	24.135	8.719	0.199	0.107	0.25	−0.195	−0.112	0.266
2.41	13.599	22.423	8.518	0.268	0.127	0.255	−0.286	−0.032	0.326
3.788	13.527	20.469	8.393	0.327	0.113	0.237	−0.307	0.037	0.349
4.996	12.995	18.409	8.361	0.376	0.084	0.195	−0.303	0.067	0.305
6.052	12.22	16.358	8.42	0.421	0.053	0.143	−0.293	0.05	0.225
6.98	11.283	14.332	8.535	0.466	0.017	0.09	−0.283	0.018	0.148
7.813	10.256	12.297	8.66	0.514	−0.012	0.046	−0.28	−0.006	0.077
8.585	9.193	10.229	8.753	0.566	−0.047	0.009	−0.288	−0.033	0.015
9.339	8.152	8.136	8.8	0.623	−0.085	−0.019	−0.312	−0.065	−0.04
10.096	7.174	6.069	8.815	0.689	−0.12	−0.049	−0.339	−0.093	−0.089
10.834	6.271	4.064	8.833	0.76	−0.155	−0.074	−0.36	−0.115	−0.123
11.537	5.442	2.153	8.883	0.839	−0.191	−0.104	−0.372	−0.131	−0.162
12.201	4.702	0.341	8.976	0.921	−0.226	−0.139	−0.381	−0.135	−0.205
12.823	4.079	−1.363	9.111	1.012	−0.257	−0.18	−0.389	−0.122	−0.25
13.407	3.597	−2.986	9.267	1.102	−0.289	−0.222	−0.395	−0.094	−0.291
13.953	3.299	−4.531	9.428	1.197	−0.314	−0.266	−0.392	−0.047	−0.328
14.444	3.222	−6.007	9.577	1.286	−0.33	−0.318	−0.355	0.02	−0.37
14.839	3.418	−7.399	9.706	1.363	−0.331	−0.378	−0.254	0.111	−0.408
15.074	3.947	−8.683	9.81	1.412	−0.312	−0.435	−0.054	0.201	−0.423
15.056	4.855	−9.837	9.878	1.422	−0.278	−0.474	0.283	0.266	−0.394
14.64	6.2	−10.814	9.893	1.382	−0.231	−0.491	0.777	0.295	−0.318

(continued)

Z. Yang et al., *Force Control Theory and Method of Human Load Carrying Exoskeleton Suit*, DOI 10.1007/978-3-662-54144-9

(continued)

Ankle joint angle	Knee joint angle	Hip joint angle	Trunk angle	Ankle joint torque	Knee joint torque	Hip joint torque	Ankle joint energy	Knee joint energy	Hip joint energy
13.627	8.084	−11.531	9.846	1.271	−0.177	−0.465	1.357	0.284	−0.194
11.765	10.621	−11.848	9.722	1.064	−0.12	−0.392	1.771	0.236	−0.05
8.847	13.945	−11.586	9.535	0.763	−0.061	−0.313	1.686	0.142	0.084
4.932	18.181	−10.581	9.304	0.443	0	−0.3	1.164	−0.074	0.294
0.46	23.361	−8.744	9.064	0.192	0.038	−0.279	0.561	−0.229	0.465
−3.808	29.341	−6.111	8.849	0.046	0.054	−0.253	0.141	−0.306	0.563
−7.001	35.748	−2.832	8.694	0	0.067	−0.219	0.016	−0.351	0.604
−8.541	42.002	0.895	8.616	−0.014	0.085	−0.2	0.016	−0.403	0.63
−8.383	47.489	4.868	8.619	−0.017	0.102	−0.182	0.025	−0.402	0.611
−6.929	51.734	8.909	8.688	−0.015	0.102	−0.146	0.03	−0.303	0.498
−4.699	54.476	12.829	8.783	−0.01	0.083	−0.096	0.024	−0.164	0.324
−2.136	55.651	16.48	8.865	−0.008	0.058	−0.043	0.018	−0.05	0.144
0.419	55.3	19.735	8.922	−0.006	0.039	−0.015	0.012	0.009	0.043
2.687	53.502	22.537	8.951	−0.006	0.024	−0.005	0.007	0.035	0.013
4.488	50.384	24.883	8.974	−0.006	0.012	−0.004	0.003	0.033	0.009
5.73	46.118	26.794	9.017	−0.007	0	0	0	−0.005	0.004
6.42	40.922	28.29	9.099	−0.009	−0.012	0.011	0	−0.08	−0.001
6.627	35.037	29.369	9.218	−0.009	−0.038	0.035	0	−0.219	−0.002
6.478	28.801	30.06	9.369	−0.009	−0.077	0.074	0	−0.413	−0.004
6.127	22.629	30.427	9.534	−0.006	−0.124	0.13	0	−0.607	−0.005
5.703	16.957	30.561	9.694	−0.003	−0.177	0.208	0	−0.739	0.002
5.303	12.215	30.551	9.831	−0.001	−0.228	0.291	0	−0.739	0.017
4.921	8.741	30.452	9.933	0	−0.253	0.343	0	−0.57	0.028
4.479	6.708	30.282	10	0	−0.247	0.342	0	−0.309	0.042
3.843	6.017	29.988	10.008	0	−0.203	0.282	0	−0.075	0.049

References

1. Wang Jimin, Li Wenxuan, Wu Shaotang, Zan Junping, Ding Aishi. The Experimental Research of Load Capacity Standard For Individual Soldier [J]. Journal of Preventive Medicine of Chinese People's Liberation Army, 1986, 4 (3): 87–88.
2. Yu Yongzhong, Li Tianlin, Liu Zunyong, Huang Haichao, Jin Shuxiang, Liu Ximei, Yang Zhenzhou, Yan Yeguang. The Research of Load Capacity Standard For Individual Soldier [J]. Chinese Journal of Industrial Hygiene and Occupational Diseases, 1991, 9 (1): 16–22.
3. Yin Zhaoyun, Zan Junping, Sun Xingbin, Ma ZhWenkao, Wang Hexin.The Relationship Between the Appropriate Load and the Slope At an Altitude of 3700 Meters Plateau [J]. Space Medicine & Medical Engineering, 1999, 12 (3): 224–226.
4. Homayoun, Seraji. Adaptive Admittance Control: An Approach to Explicit Force Control in Compliant Motion [C]. Proceedings of IEEE International Conference on Robotics and Automation, San Diego, CA, USA, 1994: 2705–2712.
5. Qiao, Bing, Lu Rongjian. Impedance force control for position controlled robotic manipulators under the constraint of unknown environments [J]. Journal of Southeast University (English Edition), 2003, 19 (4): 359–363.
6. Jaydeep, Roy, Louis L. Whitcomb. Adaptive Force Control of Position/Velocity Controlled Robots: Theory and Experiment [J]. IEEE Transactions on Robotics and Automation, 2002, 18 (2): 121–137.
7. Seraji, H. Adaptive Addittance Control: an Approach to Expilicit Force Control in Compliant Motion [C]. IEEE International Conference on Robotics and Automation, 1994: 2705–2712.
8. Pan, Zengxi. Intelligent Robotic Machining with Force Control [D]. Stevens Institute of Technology, Ph. D, Hoboken New Jercy: 2005.
9. Kuang, Yow Lian, Chia-Ru Lin. Sliding-Mode Motion/Force Control of Constrained Robots [J]. IEEE Transactions on Automatic Control, 1998, 43 (8): 1101–1103.
10. Vishnu, Mallapragada, Duygun Erol, Nilanjan Sarkar. A New Method of Force Control for Unknown Environments [C]. Proceedings of the 2006 IEEE/RSJ International Conference on Intelligent Robots and Systems, Beijing, China, 2006: 4509–4514.
11. Tsuneo, Yoshikawa. Force Control of Robot Manipulators [C]. Proceedings of the 2000 IEEE International Conference on Robotics and Automation, San Francisco, CA, USA, 2000: 220–226.
12. Chen Feng, Fei Yanqiong, Zhao Xifang. The Impedance Control Methods for Robots [J]. Control and Detection, 2005, (12): 46–48.
13. Seraji, Homayoun, Richard Colbaugh. Force Tracking in Impedance Control [C]. 1993: 499–506.
14. Mohammad, Matinfar. Optimization-Based Robot Impedance Control [D]. Queen's Universtiy, Master of Science, Kingston, Ontario, Canada: 2004.

Z. Yang et al., *Force Control Theory and Method of Human Load Carrying Exoskeleton Suit*, DOI 10.1007/978-3-662-54144-9

15. Miomir, K. Vukobratovic, Aleksandar G. Rodic, Yury Ekalo. Impedance Control as a Particular Case of the Unified Approach to the Control of Robots Interacting with a Dynamic Known Environment [J]. Journal of Intelligent and Robotics Systems, 1997, 18 (1): 191–204.
16. Seul, Jung, T. C. Hsia, Robert G. Bonitz. Force Tracking Impedance Control of Robot Manipulators Under Unknown Environment [J]. IEEE Transactions on Control Systems Technology, 2004, 12 (3): 474–483.
17. Li Jie, Wei Qing, Chang Wensen, Zhang Peng. Adaptive Force Tracking in Impedace Control [J]. Robot, 1999, 21 (1): 23–29.
18. Liu Hongyi, Wang Lei, Wang Fei. Constrained Manipulator Fuzzy Force Control Method Based on the Impedance Model in Unknown Environments [J]. Journal of Northeastern University(Natural Science), 2005, 26 (8): 767–770.
19. Han Jingqing, Wang Wei. Nonlinear Tracking Differentiator [J]. Journal of Systems Science and Mathematical Sciences, 1994, 14 (2): 177–183.
20. Han Jingqing, Yuan Lulin. The Discrete Form of Tracking Differentiator [J]. Journal of Systems Science and Mathematical Sciences, 1999, 19 (6): 268–273.

CPSIA information can be obtained
at www.ICGtesting.com
Printed in the USA
LVOW01*1740270417
532433LV00002B/13/P

9 783662 541425